Springer Proceedings in Advanced Robotics 7

The Springer Proceedings in Advanced Robotics (SPAR) publishes new developments and advances in the fields of robotics research, rapidly and informally but with a high quality.

The intent is to cover all the technical contents, applications, and multidisciplinary aspects of robotics, embedded in the fields of Mechanical Engineering, Computer Science, Electrical Engineering, Mechatronics, Control, and Life Sciences, as well as the methodologies behind them.

The publications within the "Springer Proceedings in Advanced Robotics" are primarily proceedings and post-proceedings of important conferences, symposia and congresses. They cover significant recent developments in the field, both of a foundational and applicable character. Also considered for publication are edited monographs, contributed volumes and lecture notes of exceptionally high quality and interest.

An important characteristic feature of the series is the short publication time and world-wide distribution. This permits a rapid and broad dissemination of research results.

More information about this series at http://www.springer.com/series/15556

Fanny Ficuciello · Fabio Ruggiero
Alberto Finzi
Editors

Human Friendly Robotics

10th International Workshop

 Springer

Editors
Fanny Ficuciello
Dipartimento di Ingegneria Elettrica
  e Tecnologie dell'Informazione
Università degli Studi di Napoli Federico II
Napoli
Italy

Alberto Finzi
Dipartimento di Ingegneria Elettrica
  e Tecnologie dell'Informazione
Università degli Studi di Napoli Federico II
Napoli
Italy

Fabio Ruggiero
Dipartimento di Ingegneria Elettrica
  e Tecnologie dell'Informazione
Università degli Studi di Napoli Federico II
Napoli
Italy

ISSN 2511-1256   ISSN 2511-1264   (electronic)
Springer Proceedings in Advanced Robotics
ISBN 978-3-030-07741-9   ISBN 978-3-319-89327-3   (eBook)
https://doi.org/10.1007/978-3-319-89327-3

Foreword

Robots! Robots on Mars and in oceans, in hospitals and homes, in factories and schools; robots fighting fires, making goods and products, saving time and lives. Robots today are making a considerable impact from industrial manufacturing to health care, transportation, and exploration of the deep space and sea. Tomorrow, robots will become pervasive and touch upon many aspects of modern life.

The *Springer Tracts in Advanced Robotics* (*STAR*) was launched in 2002 with the goal of bringing to the research community the latest advances in the robotics field based on their significance and quality. During the latest fifteen years, the STAR series has featured publication of both monographs and edited collections. Among the latter, the proceedings of thematic symposia devoted to excellence in robotics research, such as ISRR, ISER, FSR, and WAFR, have been regularly included in STAR.

The expansion of our field, as well as the emergence of new research areas, has motivated us to enlarge the pool of proceedings in the STAR series in the past few years. This has ultimately led to launching a sister series in parallel to STAR. The *Springer Proceedings in Advanced Robotics* (*SPAR*) is dedicated to the timely dissemination of the latest research results presented in selected symposia and workshops.

This volume of the SPAR series brings a peer-reviewed selection of the papers presented at the 10th International Workshop on Human Friendly Robotics (HFR) which took place in Naples, Italy, from November 6 to 7, 2017. HFR is an annual meeting which is organized by young researchers in a rotating fashion around Europe. The volume edited by Fanny Ficuciello, Fabio Ruggiero, and Alberto Finzi contains 16 scientific contributions ranging from physical to cognitive human–robot interaction, from grasping to manipulation, from redundant to cooperative robots, from bipedal to wearable robots.

From its excellent technical program to its warm social interaction, HFR culminates with this unique reference on the current developments and new advances in human friendly robotics—a genuine tribute to its contributors and organizers!

Napoli, Italy   Bruno Siciliano
Stanford, CA, USA   Oussama Khatib
January 2018   SPAR Editors

Preface

The growing need to automate daily tasks, combined with new robot technologies, is driving the development of a new generation of human friendly robots, i.e., safe and dependable machines, operating in the close vicinity to humans or directly interacting with them in a wide range of domains. The technological shift from classical industrial robots, which are safely kept away from humans in cages to robots, which are used in close collaboration with humans, is facing major challenges that need to be overcome. The *International Workshop on Human Friendly Robotics* (HFR) is an annual meeting dedicated to these issues and organized by young researchers in a rotating fashion around Europe. Previous venues were Naples (Italy), Genova (Italy), Pisa (Italy), Tübingen (Germany), Twente (the Netherlands), Brussels (Belgium), Rome (Italy), Pontedera (Italy), Munich (Germany), Genova (Italy). The workshop covers a wide range of topics related to human–robot interaction, both physical and cognitive, including theories, methodologies, technologies, empirical and experimental studies. The objective of the workshop is to bring together academic scientists, researchers, and research scholars to exchange and share their experiences and research results on all aspects related to the introduction of robots into everyday life. Senior scientists experts in the field are invited to the workshop as keynote speakers. Their role is also to guide students and young researchers during the discussions. The 10th International Workshop on Human Friendly Robotics (HFR 2017) was held in Naples, Italy, from November 6 to 7, 2017. The workshop was chaired by Fanny Ficuciello and co-chaired by Fabio Ruggiero and Alberto Finzi. The meeting consisted of three keynote talks, a forum with open discussion, and twenty-eight contributed presentations in a single track. This is the first edition of the workshop with associated proceedings and thus the first book of the SPAR series dedicated to HFR. The papers contained in the book have been selected on the basis of a peer-reviewed process and describe the newest and most original achievements in the field of human–robot interaction coming from the work and ideas of young researchers. Each paper presented to the workshop was refereed by at least two members of the Steering Committee.

This was composed of the following individuals:

Salvatore Anzalone	Université Paris 8
Arash Ajoudani	Istituto Italiano di Tecnologia
Matteo Bianchi	Università di Pisa
Jonathan Cacace	Università di Napoli Federico II
Riccardo Caccavale	Università di Napoli Federico II
Raffaella Carloni	University of Twente
Manuel Giuseppe Catalano	Istituto Italiano di Tecnologia
Amedeo Cesta	ISTC-CNR
Fei Chen	Advanced Robotics Department
Pietro Falco	Technical University of Munich
Alberto Finzi	Università di Napoli Federico II
Fanny Ficuciello	Università di Napoli Federico II
Manolo Garabini	Università di Pisa
Giorgio Grioli	Università di Pisa
Sami Haddadin	Leibniz Universität Hannover
Serena Ivaldi	INRIA
Dongheui Lee	Technical University of Munich
Fulvio Mastrogiovanni	Università di Genova
Andrea Orlandini	ISTC-CNR
Gianluca Palli	University of Bologna
Paolo Robuffo Giordano	IRISA, INRIA Rennes
Silvia Rossi	Università di Napoli Federico II
Fabio Ruggiero	Università di Napoli Federico II
Lorenzo Sabattini	Università di Modena and Reggio Emilia
Mario Selvaggio	Università di Napoli Federico II
Matteo Saveriano	Technical University of Munich
Bram Vanderborght	Vrije Universiteit Brussel
Valeria Villani	Università di Modena and Reggio Emilia
Andrea Maria Zanchettin	Politecnico di Milano
Chenguang Yang	Swansea University

The aim is to inaugurate a series of Springer Proceedings in Advanced Robotics (SPAR) dedicated to human friendly robotics, in order to bring, in a timely fashion, the latest advances and developments in human–robot interaction on the basis of their significance and quality. It is our hope that the wider dissemination of research developments will stimulate exchanges and collaborations among young research community and contribute to the HFR Workshop growing and expansion outside Europe. We are grateful to the authors for their contributions and to the large team of reviewers for their critical and insightful recommendations. We are proud of the three keynote talks by Prof. Oussama Khatib from University of Stanford, Dr. Rachid Alami from CNRS-LAAS, and Prof. Ciro Natale from

Università della Campania Luigi Vanvitelli. We are also indebted to Dr. Chenguang Yang and Fei Cheng for their evaluable technical contribution and to the staff of Springer who were responsible for putting the whole book together.

Napoli, Italy   Fanny Ficuciello
December 2017   Fabio Ruggiero
                                                                                          Alberto Finzi

Organization

The Human Friendly Robotics 2017 is organized by:

Organizers

Workshop Chair
Fanny Ficuciello, Università di Napoli Federico II

Workshop Co-chairs
Alberto Finzi, Università di Napoli Federico II
Fabio Ruggiero, Università di Napoli Federico II

Special Session Chairs
Fei Chen, IIT, Italy
Chenguang Yang, Swansea University, UK

Local Arrangement Chairs
Jonathan Cacace, Università di Napoli Federico II
Mario Selvaggio, Università di Napoli Federico II

Program Committee

Salvatore Anzalone, Université Paris 8
Arash Ajoudani, Istituto Italiano di Tecnologia
Matteo Bianchi, Università di Pisa
Jonathan Cacace, Università di Napoli Federico II
Riccardo Caccavale, Università di Napoli Federico II
Raffaella Carloni, University of Twente
Manuel Giuseppe Catalano, Istituto Italiano di Tecnologia

Amedeo Cesta, ISTC-CNR
Fei Chen, Advanced Robotics Department
Pietro Falco, Technical University of Munich
Alberto Finzi, Università di Napoli Federico II
Fanny Ficuciello, Università di Napoli Federico II
Manolo Garabini, Università di Pisa
Giorgio Grioli, Università di Pisa
Sami Haddadin, Leibniz Universität Hannover
Serena Ivaldi, INRIA
Dongheui Lee, Technical University of Munich
Fulvio Mastrogiovanni, Università di Genova
Andrea Orlandini, ISTC-CNR
Gianluca Palli, University of Bologna
Paolo Robuffo Giordano, IRISA, INRIA Rennes
Silvia Rossi, Università di Napoli Federico II
Fabio Ruggiero, Università di Napoli Federico II
Lorenzo Sabattini, Università di Modena and Reggio Emilia
Mario Selvaggio, Università di Napoli Federico II
Matteo Saveriano, Technical University of Munich
Bram Vanderborght, Vrije Universiteit Brussel
Valeria Villani, Università di Modena and Reggio Emilia
Andrea Maria Zanchettin, Politecnico di Milano
Chenguang Yang, Swansea University

Sponsoring Institutions

ICAROS center (Interdepartmental Center for Advances in Robotic Surgery)
CREATE consortium (Consorzio di Ricerca per l' Energia, l' Automazione e le
Tecnologie dell' Elettromagnetismo)
DIETI (Dipartimento di Ingegneria Elettrica e Tecnologie dell'Informazione)

Contents

Part I
Robot Control and Manipulation

A Stiffness-Fault-Tolerant Control Strategy for Reliable Physical Human-Robot Interaction

**Florian Stuhlenmiller, Gernot Perner, Stephan Rinderknecht
and Philipp Beckerle**

Abstract Elastic actuators allow to specify the characteristics of physical human-robot interactions and increase the intrinsic safety for the human. To ensure the reliability of the interaction, this paper investigates detection and compensation of stiffness faults. A recursive least squares algorithms is used to detect faults and obtain an estimation of the actual stiffness value online. An adaptation law based on the estimation is proposed to adjust parameters of an impedance control to maintain a desired interaction stiffness. A simulation of an exemplary elastic actuator shows that the developed stiffness-fault-tolerant control strategy achieves a dependable human-robot interaction.

Keywords Physical human-robot interaction · Fault detection · Fault-tolerant control · Safety and reliability

1 Introduction

Steady technological progress drives robots collaborating in close contact with humans [1]. An arising challenge is to provide a safe and dependable human-robot interaction [2]. Introducing elasticity to robotic joints by using series elastic actuators (SEA) facilitates back-driveability of the link and can improve inherent safety by allowing elastic deformations in presence of unknown contacts [3, 4]. Furthermore, designing or controlling the compliance of the drive train provides the possibility to adjust the physical Human-Robot Interaction (pHRI) [2, 3] online. In addition, elastic actuators are capable of storing and releasing energy in periodic trajectories, reducing energy consumption and required peak power. Therefore, natural frequencies or antiresonances of the elastic actuation system are tuned to the frequencies of the desired motion [5].

F. Stuhlenmiller (✉) · G. Perner · S. Rinderknecht · P. Beckerle
Institute for Mechatronic Systems in Mechanical Engineering,
Technical University Darmstadt, Darmstadt, Germany
e-mail: stuhlenmiller@ims.tu-darmstadt.de

© Springer International Publishing AG, part of Springer Nature 2019
F. Ficuciello et al. (eds.), *Human Friendly Robotics*, Springer Proceedings
in Advanced Robotics 7, https://doi.org/10.1007/978-3-319-89327-3_1

3

While elastic actuators exhibit many advantages, the complexity of the drive train can lead to an increased fault-sensitivity and reduced reliability. According to an expert study, faults have a high practical relevance for the utilization of elastic actuators [6]. Faults occur thereby frequently in kinematic components, sensors and software. However, elastic elements are rated with increased probability ratings, which supports their practical relevance [6]. Especially in close contact between human and robot, faults of the elastic robotic system can compromise safety, hence, a fault management system appears crucial to ensure a safe and reliable pHRI [7].

A fault is generally defined as an inadmissible deviation of a property from the desired condition [8], which leads to a system that does not fulfill its function properly [9]. Fault diagnosis is required to detect deviations, identify the malfunctioning component, and perform countermeasures, e.g. transfer the system into a safe operational state or enable specific modes of operation by means of fault-tolerant control [9].

This paper focuses on the detection, identification, and compensation of faults occurring in the elastic element, which is an unique component of elastic actuators and has a significant influence on pHRI. Therefore, a fault-tolerant control strategy is designed and applied to a series elastic actuator in Sect. 2. The model-based detection of a stiffness fault and corresponding compensation strategy is developed in Sect. 3. Simulation results in Sect. 4 show the feasibility of the proposed method. At last, a conclusion is given in Sect. 5.

2   Actuator and Control

In this section, an actuator with Variable Torsional Stiffness (VTS) [10, 11] is presented. It serves as an exemplary SEA to develop and evaluate the dependability of pHRI in presence of a stiffness fault and develop appropriate countermeasures. Further, an introduction to impedance control is given since it forms the basis of the proposed stiffness-fault-tolerant control strategy.

2.1   Actuator Model

The VTS-actuator shown in Fig. 1 consists of an elastic element in series to actuator 1 and drives a pendulum load. Actuator 2 moves a counter bearing on the elastic element to control the stiffness k_s by adjusting the active length of the spring. This mechanism can also be used to emulate a fault by setting deviating stiffness values.

Generally, SEA-driven joints can be represented by models with two degrees of freedom. As can be seen in Fig. 2, actuator inertia I_a and link inertia I_l are coupled by the series stiffness k_s. The actuator generates the torque τ_a, the link is loaded by the external torque τ_{ext}. The positions at actuator and link are given by φ_a and φ_l, respectively. Neglecting damping and centrifugal effects leads to the following system of equations:

(a) **(b)**

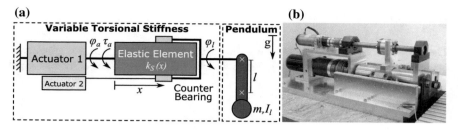

Fig. 1 Structure (left) and photography (right) of the VTS-actuator

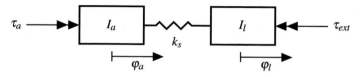

Fig. 2 General two-degree-of-freedom model of a SEA

$$\begin{bmatrix} I_l & 0 \\ 0 & I_a \end{bmatrix} \begin{bmatrix} \ddot{\varphi}_l \\ \ddot{\varphi}_a \end{bmatrix} + \begin{bmatrix} k_s & -k_s \\ -k_s & k_s \end{bmatrix} \begin{bmatrix} \varphi_l \\ \varphi_a \end{bmatrix} = \begin{bmatrix} \tau_{ext} \\ \tau_a \end{bmatrix} \tag{1}$$

For VTS, the external loads consist of the gravitational torque of the pendulum, determined by the mass m, the distance to the center of rotation l and the gravitational acceleration g as well as the an external disturbance τ_{dis}:

$$\tau_{ext} = -mlg \sin(\varphi_l) + \tau_{dis} \tag{2}$$

In the following, all loads due to pHRI are considered as external disturbance and described by τ_{dis}. The parameters of the VTS-actuator according to the identification in [12] are given in Table 1.

Table 1 Parameters of the VTS-prototype

Description		Value
Inertia of actuator	I_a	$1.15\,\mathrm{kg\,m^2}$
Inertia of pendulum	I_l	$0.9\,\mathrm{kg\,m^2}$
Mass of pendulum	m	$6.81\,\mathrm{kg}$
Length of pendulum	l	$0.36\,\mathrm{m}$
Gravitational acceleration	g	$9.81\,\mathrm{ms^{-2}}$
Torsional stiffness	k_s	50–$350\,\mathrm{N\,m\,rad^{-1}}$ (variable)

2.2   *Impedance Control*

The proposed fault-tolerant control strategy bases on passivity-based impedance controller as described in [13]. To shape the inertial properties, the actuator torque is determined from:

$$\tau_a = \frac{I_a}{I_{a,d}} u + \left(1 - \frac{I_a}{I_{a,d}}\right) k_s (\varphi_a - \varphi_l) \tag{3}$$

The control input u is determined to achieve passivity of the system [13], yielding:

$$u = I_{a,d} \ddot{\varphi}_{a,d} + k_s (\varphi_{a,d} - \varphi_{l,d}) + \\ + k_c (\varphi_{a,d} - \varphi_a) + d_c (\dot{\varphi}_{a,d} - \dot{\varphi}_a) \tag{4}$$

To track a motion, the control input considers the desired position of the link $\varphi_{l,d}$ and the corresponding gravitational torque $mlg \sin(\varphi_{l,d})$ to calculate the desired actuator angle $\varphi_{a,d}$ via inverse dynamics from Eq. (1). For the VTs-actuator, this yields:

$$\varphi_{a,d} = \varphi_{l,d} + \frac{1}{k_s} \left(mlg \sin(\varphi_{l,d}) + I_l \ddot{\varphi}_{l,d}\right) \tag{5}$$

The first two terms of Eq. (4) represent compensation of loads for the desired motion of link and actuator. The last two terms in Eq. (4) represent PD-control of the actuator and define the reaction to external, unknown disturbances. The PD-controller implements a virtual stiffness k_c and viscous damping d_c with respect to the actuator, yielding the characteristic structure of the impedance-controlled SEA presented in Fig. 3. This leads to the following system dynamics with respect to the position error $\tilde{\varphi}$:

$$\begin{bmatrix} I_l & 0 \\ 0 & I_{a,d} \end{bmatrix} \begin{bmatrix} \ddot{\tilde{\varphi}}_l \\ \ddot{\tilde{\varphi}}_a \end{bmatrix} + \begin{bmatrix} 0 & 0 \\ 0 & d_c \end{bmatrix} \begin{bmatrix} \dot{\tilde{\varphi}}_l \\ \dot{\tilde{\varphi}}_a \end{bmatrix} + \begin{bmatrix} k_s & -k_s \\ -k_s & k_s + k_c \end{bmatrix} \begin{bmatrix} \tilde{\varphi}_l \\ \tilde{\varphi}_a \end{bmatrix} = \begin{bmatrix} \tau_{dis} \\ 0 \end{bmatrix} \tag{6}$$

Hence, the interaction stiffness k_i as the relation between the disturbance τ_{dis} and position error at the link $\tilde{\varphi}_l$ is a series configuration of virtual and physical stiffness:

$$k_i = \frac{k_s k_c}{k_s + k_c} \tag{7}$$

Modeling the pHRI as external disturbance, the impedance control allows to set the interaction stiffness k_i between actuator and human via the control parameter k_c. Configuring k_c generates the possibility to create a situational or user-specific characteristic of the pHRI.

However, to achieve a specific stiffness of the interaction, detailed knowledge about the value of the physical stiffness k_s is required. Furthermore, k_s is required as input to determine the desired motion of the actuator with Eq. (5) as well as the impedance control law given by Eqs. (3) and (4). Hence, the occurrence of a fault,

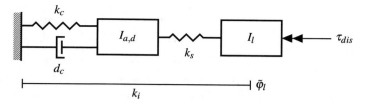

Fig. 3 Model of an impedance-controlled SEA

which causes a change of the physical stiffness, would reduce control performance and influence pHRI characteristics.

3 Detection and Adaptation to Fault

This section presents the detection of a stiffness fault and an adaptation of the control strategy in order to compensate for the malfunction and to maintain specified pHIR characteristics.

3.1 Control Strategy

The control strategy presented in Fig. 4 adapts the stiffness parameter k_c of the impedance control algorithm to compensate for deviations of the physical stiffness k_s. Therefore, k_c is set according to an estimation of the physical stiffness $\bar{k}_s$ to retain the desired pHRI characteristics. The upper part of Fig. 4 shows the presented impedance control, composed of inverse dynamics (Eq. 5), the impedance control law (Eqs. 3 and 4) and the plant (Eq. 1). An estimation of the physical stiffness $\bar{k}_s (\tau_a, \varphi_a, \dot{\varphi}_a, \varphi_l)$ is determined based on the approaches given in [14–16], fed back to the inverse dynamics and impedance control, and used to determine k_c.

3.2 Stiffness Estimation

To detect faults during operation, online-applicability of the fault detection and adaptation is required. Reliable stiffness estimation with high convergence rates is required to achieve the desired control behavior. Therefore, a recursive stiffness estimation algorithm that consists of two consecutive calculations is adopted [14–16]. This algorithm consists of two consecutive calculations. First, model-based determination and low-pass filtering of the torque in the elastic element is performed to be robust against outliers. This estimated value is passed to a recursive least squares

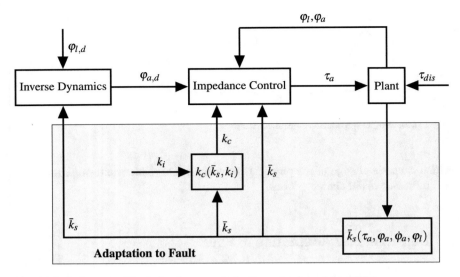

Fig. 4 Block diagram of the fault-tolerant control strategy for dependable pHRI

(RLS) algorithm with fixed directional forgetting [17] to dynamically compute the estimated physical stiffness $\bar{k}_s$. To estimate and filter the spring torque, an equivalent residual r_e is generated from the part of the model that represents the actuator, i.e., the lower line of Eq. (1). With a positive design parameter K, the residual is calculated from:

$$\bar{\tau}_e \approx r_e = K \left(I_a \dot{\varphi}_a + D \varphi_a - \int_0^{t_1} (\tau_a + r_e) \, dt \right). \tag{8}$$

As this continuous time expression is not applicable to discrete-time control, it is formulated as a recursive algorithm according based on Tustin rule [18]. For the time step j and sampling time T, the residual $r_e(j)$ is calculated by

$$J(j) = J(j-1) + \frac{\tau_a(j) + \tau_a(j-1)}{2} T \tag{9}$$

$$r(j) = K \left(I_a \dot{\varphi}_a(j) + D \varphi_a(j) - J(j) \right) \tag{10}$$

$$r_e(j) = \frac{2 - TK}{2 + TK} r_e(j-1) + \frac{2(r(j) - r(j-1))}{2 + TK} \tag{11}$$

For estimation of the physical stiffness, an RLS algorithm with directional forgetting is proposed that uses a linear parameterization of the unknown stiffness parameters α_h based on a parametric model $f(\phi, \boldsymbol{\alpha})$:

$$f(\Delta \varphi, \boldsymbol{\alpha}) = \sum_{h=1}^{n} f_h(\Delta \varphi) \alpha_h. \tag{12}$$

The function f only depends on the deflection of the spring $\phi = \varphi_l - \varphi_a$ of the elastic element. Under the assumption of a linear elastic behavior, Eq. 12 simplifies to

$$\bar{\tau}_e \approx r_e = \bar{k}_s \phi. \tag{13}$$

Based on the recursive calculation of the residual r_e in Eqs. (9) to (11) and the linear model for elasticity from Eq. 13 an on-line estimation of $\bar{k}_s$ is performed with a discrete-time recursive least squares algorithm with directional forgetting. Hereby, the input signals are ϕ as regressor and r_e as response.

$$\varepsilon(j-1) = \lambda - \frac{1-\lambda}{\phi(j)C(j-1)\phi(j)} \tag{14}$$

$$C(j) = C(j-1) - \frac{C(j-1)\phi(j)\phi(j)C(j-1)}{\varepsilon(j-1)^{-1} + \phi(j)C(j-1)\phi(j)} \tag{15}$$

$$L(j) = \frac{C(j-1)\phi(j)}{1 + (\phi(j) \cdot C(j-1) \cdot \phi(j))} \tag{16}$$

$$\bar{k}_s(j) = \bar{k}_s(j-1) + L(j) \tag{17}$$

In this set of equations, $C(j)$ is the covariance matrix of the estimated parameter $\bar{k}_s(j)$ and $L(j)$ denotes the gain matrix. The directional forgetting factor $\varepsilon(j-1)$ is adapted by the forgetting factor $0 < \lambda < 1$. In contrast to the standard RLS algorithm, this algorithm is more suitable for dynamically changing parameters and avoids a wind-up of the estimation by introducing the directional forgetting factor $\varepsilon(j-1)$ [17].

3.3 Adaptation of Control Parameters

Inserting the estimated physical stiffness $\bar{k}_s$ into Eq. 7 and rearranging to determine the virtual stiffness k_c to achieve a given, desired interaction stiffness k_i yields the adaptation law:

$$k_c = \frac{\bar{k}_s k_i}{\bar{k}_s - k_i} \tag{18}$$

By adapting the control parameter k_c to the resulting value, fault-tolerant pHRI is ensured. With the physical representation of an impedance controlled SEA given in Fig. 3, the stability of the system can be analyzed based on passivity properties of the system. According to [13],

$$k_c > 0 \tag{19}$$

is required to achieve a passive system. It follows from Eq. (7) that

$$k_i < k_s. \tag{20}$$

This condition for stability limits the range of adaptation of k_i, as the pHRI can only be softer than the physical stiffness. Additionally, the selection of high gains due to high desired values of k_c can practically limit control performance.

4   Simulation Results

The reaction of the system to faults occurring in the physical stiffness as well as the feasibility of the proposed method is evaluated in simulation. To evaluate stiffness estimation and control adaptation, dynamic behavior of the VTS-actuator in interaction with the human during fault occurrence is considered. A fault of the stiffness is modeled as a sigmoid progression of k_{fault}, characterized by a decrease of p_{decr} in % of k_s at time t_{fault} according to:

$$k_{fault}(t) = k_s \left(1 - p_{decr}\left(1 + e^{-a(t-t_{fault})}\right)^{-1}\right) \tag{21}$$

Thereby, a affects the gradient of transition of the sigmoid function and is set to $a = 20\,s^{-1}$.

As a simple case of pHRI, a contact with a human is considered. It is represented by a spring-damper system with the stiffness $k_h = 500\,\mathrm{N\,m\,rad^{-1}}$ and the damping coefficient $d_h = 20\,\mathrm{N\,m\,rad^{-1}}$ as depicted in Fig. 5. The disturbance is activated for positive angles:

$$\tau_{dis} = \begin{cases} 0 & \text{for} \quad \varphi_l < 0 \\ k_h(\varphi_l - \varphi_{dis}) + d_h\dot{\varphi}_l & \text{for} \quad \varphi_l \geq 0 \end{cases} \tag{22}$$

Hence, in this case, the interaction stiffness k_i determines the contact force and how far the point of contact is pushed to negative φ_{dis} values by the pendulum. While in general, compliance and damping of the human are non-linear and comprised with

Fig. 5 Abstraction of the interaction between link and human

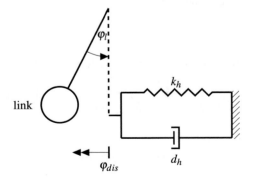

uncertainties, the simplified model is deemed sufficient to show that the presented fault-tolerant control strategy achieves the desired interaction stiffness.

A simulation is performed for a sinusoidal motion of the link with an amplitude of $10°$, a frequency of $0.5\,\text{Hz}$, and a desired interaction stiffness of $k_i = 50\,\text{N m rad}^{-1}$. The pHRI takes place between $t = 10$ and $18\,\text{s}$ to provide sufficient time for reaching steady-state behavior. Considering the interval $10\,\text{s} < t < 15\,\text{s}$ allows to assess the estimation and control performance in nominal operation, while the effects due to fault occurrence are analyzed for $t \geq 15\,\text{s}$. The control parameter k_c is directly determined from Eq. 18, d_c is set to $20\,\text{N m rad}^{-1}$. The desired actuator inertia is not changed from the original value, i.e., $I_{a,d} = I_a$. In the residual generator, K_{τ_a} is iteratively tuned to a value of 500. For the stiffness estimation, an exponential forgetting factor of 0.999 is selected. The stiffness and the covariance matrix are initialized by $0.8k_s$ and with a unity matrix scaled with the factor 10, respectively.

4.1 Stiffness Estimation and Adaptation of Control Parameters

The physical stiffness is set to $k_s = 100\,\text{N m rad}^{-1}$ and reduced by $p_{decr} = 30\%$ to $70\,\text{N m rad}^{-1}$ at $t_{fault} = 15\,\text{s}$ as presented in Fig. 6. During steady-state operation before fault occurrence, the estimated stiffness coincides with the real one. After the stiffness fault occurs, a period of approximately $2\,\text{s}$ is required until the estimation converges to the new stiffness value, which is deemed fast enough to compensate for a stiffness fault after one oscillation. The mean squared error between estimated stiffness and true value is $0.35\,\text{N m rad}^{-1}$, thus, the estimation is sufficiently accurate for fault-compensation.

A simulation without compensation of the fault is presented in Fig. 7. The disturbance τ_{dis} shown in the upper plot results from pHRI. The compliant behavior of the system can be observed in the middle and lower plots which show the deviation of the link position. The desired system behavior (red) is shown as reference and is determined from τ_{dis} via Eq. 6. After the occurrence of the stiffness fault, which is indicated by the dashed black line, the compliance of the system is higher

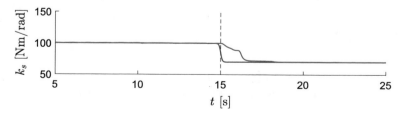

Fig. 6 Progression of the real stiffness (blue) and the estimated stiffness (red) with $t_{fault} = 15\,\text{s}$ (dashed-black)

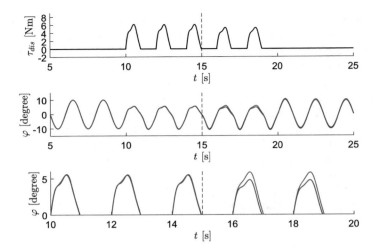

Fig. 7 Simulation without compensation of the stiffness fault with $t_{fault} = 15\,$s (dashed-black); top: interaction torque (black), middle: position link (blue), desired position link (red); bottom: magnification of the plot in the middle

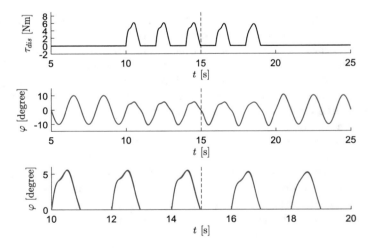

Fig. 8 Simulation with compensation of the stiffness fault with $t_{fault} = 15\,$s (dashed-black); top: interaction torque (black), middle: position link (blue), desired position link (red); bottom: magnification of the plot in the middle

than desired, and the link position deviates from the desired trajectory. This also results in lower interaction force. Furthermore, the control performance decreases even without disturbance.

Figure 8 shows the results obtained with activated stiffness estimation and corresponding control parameter adaptation. The link tracks the desired trajectory with negligible position error and the system exhibits the specified interaction behavior as red and blue curve are overlapping throughout the simulation. Hence, the stiffness-fault is compensated and a fault-tolerant pHRI is achieved.

5 Conclusion

This paper presents a method to detect a stiffness fault of SEAs via online estimation of the physical stiffness to provide reliable pHRI by impedance control adaptation. Stiffness estimations are determined via a model-based calculation of the spring torque and a recursive least squares algorithm. An observation of variations in the estimation allows to detect stiffness faults. In order to compensate for a change in the compliance and maintain the desired interaction stiffness, the estimated physical stiffness is used to adapt the virtual stiffness of the impedance control. Simulations considering a stiffness fault in an actuator with variable torsion stiffness point out that the accuracy, convergence, and robustness of the stiffness estimation are suitable. With the presented fault-tolerant control approach, a reliable behavior during faulty operating states and a preservation of the desired interaction characteristics is ensured. By enabling fault-tolerant pHRI, the proposed control strategy contributes to safe system behavior, which should to be investigated experimentally in future work.

Acknowledgements This work was supported by a Deutsche Forschungsgemeinschaft (DFG) Research Grant (no. BE 5729/1-1).

References

1. Vasic, M., Billard, A.: Safety issues in human-robot interactions. In: 2013 IEEE International Conference on Robotics and Automation (ICRA), pp. 197–204. IEEE (2013)
2. Haddadin, S., Albu-Schffer, A., Hirzinger, G.: Safe physical human-robot interaction: measurements, analysis and new insights. In: ISRR, vol. 66, pp. 395–407. Springer (2007)
3. de Santis, A., Siciliano, B., de Luca, A., Bicchi, A.: An atlas of physical human robot interaction. Mechan Mach Theory **43**(3), 253–270 (2008)
4. Bicchi, A., Bavaro, M., Boccadamo, G., De Carli, D., Filippini, R., Grioli, G., Piccigallo, M., Rosi, A., Schiavi, R., Sen, S.: Others, physical human-robot interaction: dependability, safety, and performance. In: 10th IEEE International Workshop on Advanced Motion Control: AMC'08, vol. 2008, pp. 9–14. IEEE (2008)
5. Verstraten, T., Beckerle, P., Furnmont, R., Mathijssen, G., Vanderborght, B., Lefeber, D.: Series and parallel elastic actuation: impact of natural dynamics on power and energy consumption. Mechan Mach Theory **102**, 232–246 (2016)
6. Beckerle, P.: Practical relevance of faults, diagnosis methods, and tolerance measures in elastically actuated robots. Control Eng. Pract **50**, 95–100 (2016)
7. Filippini, R., Sen, S., Bicchi, A.: Toward soft robots you can depend on. IEEE Robot. Autom. Mag. **15**(3), 31–41 (2008)
8. Isermann, R.: Fault-Diagnosis Systems: an Introduction from Fault Detection to Fault Tolerance. Springer (2006)

9. Blanke, M., Kinnaert, M., Lunze, J., Staroswiecki, M.: Diagnosis and Fault-Tolerant Control: with 218 Figures, 129 Examples, and 43 Exercises, 3rd edn. Springer (2016)
10. Beckerle, P., Wojtusch, J., Schuy, J., Strah, B., Rinderknecht, S., von Stryk, O.: Power-optimized stiffness and nonlinear position control of an actuator with variable torsion stiffness. In: IEEE/ASME International Conference on Advanced Intelligent Mechatronics (2013)
11. Lendermann, M., Singh, B.R.P., Stuhlenmiller, F., Beckerle, P., Rinderknecht, S., Manivannan, P.V.: Comparison of passivity based impedance controllers without torque-feedback for variable stiffness actuators. In: IEEE/ASME International Conference on Advanced Intelligent Mechatronics (2015)
12. Lendermann, M., Stuhlenmiller, F., Erler, P., Beckerle, P., Rinderknecht, S.: A systematic approach to experimental modeling of elastic actuators by component-wise parameter identification. In: IEEE/RSJ International Conference on Intelligent Robots and Systems, Oct. 2015
13. Ott, C.: Cartesian Impedance Control of Redundant and Flexible-Joint Robots. Springer (2008)
14. Perner, G., Yousif, L., Rinderknecht, S., Beckerle, P.: Feature extraction for fault diagnosis in series elastic actuators. In: Conference on Control and Fault-Tolerant Systems (2016)
15. Flacco, F., de Luca, A., Sardellitti, I., Tsagarakis, N.G.: On-line estimation of variable stiffness in flexible robot joints. Int. J. Robot. Res. **31**(13), 1556–1577 (2012)
16. Flacco, F., de Luca, A.: Residual-based stiffness estimation in robots with flexible transmissions. In: 2011 IEEE International Conference on Robotics and Automation (ICRA), pp. 5541–5547. IEEE (2011)
17. Navrtil, P., Ivanka, J.: Recursive estimation algorithms in matlab & simulink development environment. WSEAS Trans. Comput. **13**, 691–702 (2014)
18. Flacco, F., de Luca, A.: Stiffness estimation and nonlinear control of robots with variable stiffness actuation. In: IFAC World Congress, pp. 6872–6879 (2011)

Tracking Control of Redundant Manipulators with Singularity-Free Orientation Representation and Null-Space Compliant Behaviour

Fabio Vigoriti, Fabio Ruggiero, Vincenzo Lippiello and Luigi Villani

Abstract This paper presents a suitable solution to control the pose of the end-effector of a redundant robot along a pre-planned trajectory, while addressing an active compliant behaviour in the null-space. The orientation of the robot is expressed through a singularity-free representation form. To accomplish the task, no exteroceptive sensor is needed. While a rigorous stability proof confirms the developed theory, experimental results bolster the performance of the proposed approach.

Keywords Redundant robot tracking · Null-space compliance · Singularity-free orientation representation

1 Introduction

Over the last years, robotic systems are started to be used in those application areas where human-robot physical interaction becomes unavoidable and necessary. A new term, Socially Assistive Robotics (SAR), is defined in [6] for service robots working with humans. In certain cases, it might also be useful to measure, or at least estimate, the exchanged forces and to figure out whether the contact with the human operator has been unintentional or intentional (i.e., required for collaborative tasks). The need of safety and dependability measures is discussed in [8] based on impact tests of a

F. Vigoriti · F. Ruggiero (✉) · V. Lippiello · L. Villani
Consorzio CREATE and PRISMA Lab, Department of Electrical Engineering
and Information Technology, Università degli Studi di Napoli Federico II, Via Claudio 21, 80125
Naples, Italy
e-mail: fabio.ruggiero@unina.it

F. Vigoriti
e-mail: fabio.vigoriti@unina.it

V. Lippiello
e-mail: vincenzo.lippiello@unina.it

L. Villani
e-mail: luigi.villani@unina.it

© Springer International Publishing AG, part of Springer Nature 2019
F. Ficuciello et al. (eds.), *Human Friendly Robotics*, Springer Proceedings
in Advanced Robotics 7, https://doi.org/10.1007/978-3-319-89327-3_2

15

lightweight robot with a crash-test dummy for possible injuries that can happen in a SAR system, and on the severity of these injuries.

Hence, for safety reasons, a compliant behaviour is often requested to a robot. Such compliance can be in principle achieved through either a mechanical device or a suitable control law. In the former case, an elastic decoupling is placed between the actuators and the link, obtaining a fixed or a variable joint stiffness [1]. In the latter case, the compliant behaviour is obtained via software, like implementing an impedance control [7, 9, 19].

In the research project PHRIDOM, see [5], the components of a robotic application are discussed based on safety and dependability in physical human robot interaction. The mentioned components are mechanics, actuation control techniques and real-time planning for safety measures.

In the event of using a redundant robot, the compliant behaviour can be obtained both at the main task level [15, 17] and in the null-space [18]. This last is helpful when the interaction control cannot interfere with the execution of the main task. In principle, the resulting external wrench affecting the main task should be properly measured or estimated: in this way, it is possible to design an impedance behaviour in the null-space without seizing the main one.

Using a redundant robot to minimize the injury possibilities prior to detect the contact is proposed in [13]. To this aim, a posture optimization technique is employed to make a redundant robot arm able to change its posture to minimize the impact forces along a given direction while carrying out the main task.

This paper extends what presented in [18, 20] by explicitly addressing the tracking case and employing a singularity-free representation for the orientation of the robot's end-effector, e.g., axis-angle or unit quaternion. Notice that using one of these two orientation representations, the theoretical framework in [18] fails.[1] The sought aim is to control the robot arm while the end-effector has to follow a pre-planned trajectory in terms of both position and orientation, and the manipulator has to exhibit an active compliant behaviour in the null-space. A rigorous stability analysis is carried out thanks to the presence of a dynamic term in the controller, filtering both the effects of the velocity and of the external wrench, while no exteroceptive sensors are needed to fulfil the given task.

2 Mathematical Framework

2.1 Notation

A redundant robot manipulator ($n > 6$) is considered in this paper, with n the number of joints. The vector $q \in \mathbb{R}^n$ denotes the joint positions, while $\dot{q} \in \mathbb{R}^n$ and $\ddot{q} \in \mathbb{R}^n$ the joint velocities and accelerations, respectively.

---

[1]In detail, the proof of Proposition 3 within [18] cannot be applied whether the orientation error is chosen as it will be defined in this paper.

Let Σ_i and Σ_e be the world inertial frame and the end-effector frame, respectively. The desired trajectory for Σ_e is specified in terms of the desired time-varying position $\boldsymbol{p}_d \in \mathbb{R}^3$, orientation $\boldsymbol{R}_d \in SO(3)$, linear velocity $\dot{\boldsymbol{p}}_d \in \mathbb{R}^3$, angular velocity $\boldsymbol{\omega}_d \in \mathbb{R}^3$, linear acceleration $\ddot{\boldsymbol{p}}_d \in \mathbb{R}^3$ and angular acceleration $\dot{\boldsymbol{\omega}}_d \in \mathbb{R}^3$, all expressed in Σ_i. On the other hand, $\boldsymbol{p}_e \in \mathbb{R}^3$ denotes the current position, $\dot{\boldsymbol{p}}_e \in \mathbb{R}^3$ the current linear velocity, $\boldsymbol{R}_e \in SO(3)$ the current orientation and $\boldsymbol{\omega}_e \in \mathbb{R}^3$ the current angular velocity of Σ_e with respect to Σ_i. The twist $\boldsymbol{v} = \begin{bmatrix} \dot{\boldsymbol{p}}_e^{\mathrm{T}} & \boldsymbol{\omega}_e^{\mathrm{T}} \end{bmatrix}^{\mathrm{T}} \in \mathbb{R}^6$ is used to compact notation. Finally, the identity and zero matrices are denoted by $\boldsymbol{I}_a \in \mathbb{R}^{a \times a}$ and $\boldsymbol{O}_a \in \mathbb{R}^{a \times a}$, respectively, while the zero vector is denoted by $\boldsymbol{0}_a \in \mathbb{R}^a$. With the given notation in mind, it is possible to define the so-called deviation matrix as $\widetilde{\boldsymbol{R}} = \boldsymbol{R}_d^{\mathrm{T}} \boldsymbol{R}_e \in SO(3)$. A non-minimal representation for $\widetilde{\boldsymbol{R}}$ can be obtained by resorting to 4 parameters $\boldsymbol{\alpha} \in \mathbb{R}^4$ [2, 19].

2.2 Axis-Angle and Unit Quaternion

The matrix $\widetilde{\boldsymbol{R}}$ can be seen as a rotation of an angle $\widetilde{\phi} \in \mathbb{R}$ around the unit vector $\widetilde{\boldsymbol{r}} \in \mathbb{R}^3$. Such representation is not unique because $\widetilde{\boldsymbol{R}}(\widetilde{\phi}, \widetilde{\boldsymbol{r}}) = \widetilde{\boldsymbol{R}}(-\widetilde{\phi}, -\widetilde{\boldsymbol{r}})$. Expressing the columns of the current $\boldsymbol{n}_e, \boldsymbol{s}_e, \boldsymbol{a}_e \in \mathbb{R}^3$ and desired $\boldsymbol{n}_d, \boldsymbol{s}_d, \boldsymbol{a}_d \in \mathbb{R}^3$ rotation matrix of Σ_e with respect to Σ_i as $\boldsymbol{R}_e = \begin{bmatrix} \boldsymbol{n}_e & \boldsymbol{s}_e & \boldsymbol{a}_e \end{bmatrix}$ and $\boldsymbol{R}_d = \begin{bmatrix} \boldsymbol{n}_d & \boldsymbol{s}_d & \boldsymbol{a}_d \end{bmatrix}$, respectively, it is possible to choose the following compact expression for the orientation error [19]: $\widetilde{\boldsymbol{o}} = \frac{1}{2} \left(S(\boldsymbol{n}_e)\,\boldsymbol{n}_d + S(\boldsymbol{s}_e)\,\boldsymbol{s}_d + S(\boldsymbol{a}_e)\,\boldsymbol{a}_d \right)$, where $S(\cdot) \in \mathbb{R}^{3 \times 3}$ is the skew-symmetric matrix. When the axes of $\boldsymbol{R}_d$ and $\boldsymbol{R}_e$ are aligned, $\widetilde{\boldsymbol{o}}$ is zero and $\widetilde{\boldsymbol{R}} = \boldsymbol{I}_3$. An equivalent expression for the orientation error in axis-angle representation is given by $\widetilde{\boldsymbol{o}} = \sin(\widetilde{\phi})\widetilde{\boldsymbol{r}}$ [2].

The ambiguity of the axis-angle representation can be overcome by introducing the *unit quaternion*. Define the quantities $\widetilde{\eta} = \cos\left(\widetilde{\phi}/2\right)$ and $\widetilde{\boldsymbol{\epsilon}} = \sin\left(\widetilde{\phi}/2\right)\widetilde{\boldsymbol{r}}$, where $\widetilde{\boldsymbol{\epsilon}} \in \mathbb{R}^3$ is the *vectorial part* of the quaternion, while $\widetilde{\eta} \in \mathbb{R}$ is its *scalar part*. The quaternion is referred to as unit since it satisfies $\widetilde{\boldsymbol{\epsilon}}^{\mathrm{T}}\widetilde{\boldsymbol{\epsilon}} + \widetilde{\eta}^2 = 1$, and it is a double cover of $SO(3)$ since it can be shown that $(\widetilde{\eta}, \widetilde{\boldsymbol{\epsilon}})$ and $(-\widetilde{\eta}, -\widetilde{\boldsymbol{\epsilon}})$ corresponds to the same rotation matrix. Therefore, $\widetilde{\boldsymbol{\epsilon}} = \boldsymbol{0}_3$ if and only if $\widetilde{\eta} = \pm 1$. This means that both $(\widetilde{\eta} = 1, \widetilde{\boldsymbol{\epsilon}} = \boldsymbol{0}_3)$ and $(\widetilde{\eta} = -1, \widetilde{\boldsymbol{\epsilon}} = \boldsymbol{0}_3)$ correspond to $\widetilde{\boldsymbol{R}} = \boldsymbol{I}_3$, and thus $\boldsymbol{R}_e = \boldsymbol{R}_d$ as desired.

The orientation error for the axis-angle representation can be expressed in terms of unit quaternion as $\widetilde{\boldsymbol{o}} = 2\widetilde{\eta}\widetilde{\boldsymbol{\epsilon}}$ [2]. In this paper the following orientation error definition is instead preferred $\widetilde{\boldsymbol{o}} = \widetilde{\boldsymbol{\epsilon}}$. It is possible to prove that, in case $\widetilde{\boldsymbol{o}} = \widetilde{\boldsymbol{\epsilon}} = \boldsymbol{0}_3$, the indetermination $\widetilde{\eta} = \pm 1$ does not affect the system stability [2]. Moreover, after deriving the kinematic equations for the unit quaternion as $\dot{\widetilde{\eta}} = (1/2)\widetilde{\boldsymbol{\epsilon}}^{\mathrm{T}}\widetilde{\boldsymbol{\omega}}$ and $\dot{\widetilde{\boldsymbol{\epsilon}}} = -(1/2)\left(\widetilde{\eta}\boldsymbol{I}_3 + S(\widetilde{\boldsymbol{\epsilon}})\right)\widetilde{\boldsymbol{\omega}}$, set $\boldsymbol{\alpha} = \begin{bmatrix} \widetilde{\eta} & \widetilde{\boldsymbol{\epsilon}}^{\mathrm{T}} \end{bmatrix}^{\mathrm{T}}$, the time derivative of the orientation error can be written as

$$\dot{\widetilde{\boldsymbol{o}}} = \boldsymbol{L}_q(\boldsymbol{\alpha})\widetilde{\boldsymbol{\omega}}, \tag{1}$$

where $\widetilde{\omega} = \omega_d - \omega_e \in \mathbb{R}^3$ is the angular velocity error and $L_q(\alpha) = -(1/2)$ $(\widetilde{\eta}I_3 + S(\widetilde{\epsilon})) \in \mathbb{R}^{3\times3}$ is a nonsingular matrix.

2.3 Dynamics

The dynamic model of a robot arm in the joint space can be written as [19]

$$B(q)\ddot{q} + C(q, \dot{q})\dot{q} + g(q) = \tau - \tau_{ext}, \tag{2}$$

where $B(q) \in \mathbb{R}^{n\times n}$ is the inertia matrix in the joint space; $C(q, \dot{q}) \in \mathbb{R}^{n\times n}$ is the so-called Coriolis matrix; $g(q) \in \mathbb{R}^n$ is the vector collecting gravity terms; $\tau \in \mathbb{R}^n$ is the control torques vector; $\tau_{ext} \in \mathbb{R}^n$ is the vector representing the effect of the resulting external wrench mapped on the joints. Since no force/torque sensors are employed, τ_{ext} cannot be measured.

Denoting with $J(q) \in \mathbb{R}^{6\times n}$ the geometric Jacobian of the robot arm, the equation $v = J(q)\dot{q}$, holds [19]. As an assumption, $J(q)$ is always full rank.[2] Following the *joint space decomposition method* [15], it is possible to add $r = n - 6$ auxiliary variables $\lambda \in \mathbb{R}^r$ to the end-effector velocity v defined as $\dot{q} = Z(q)\lambda$, where $Z(q) \in \mathbb{R}^{n\times r}$ is such that $J(q)Z(q) = O_{6\times r}$. Notice that $Z(q)$ spans the null-space of $J(q)$. A convenient choice for λ is given by the left inertia-weighted generalized inverse of $Z(q)$ [14], such that $\lambda = Z(q)^{\#}\dot{q}$, with $Z(q)^{\#} = \left(Z(q)^{\mathrm{T}}B(q)Z(q)\right)^{-1}Z(q)^{\mathrm{T}}B(q)$. In the same way, it is possible to define a dynamically consistent generalized inverse Jacobian [12] as $J(q)^{\#} = B(q)^{-1}J(q)^{\mathrm{T}}\left(J(q)B(q)^{-1}J(q)^{\mathrm{T}}\right)^{-1}$ whose metrics is induced by the inertia matrix, as well as for $Z(q)^{\#}$, and that plays a key role in null-space dynamics [15]. Therefore, the following decomposition for the joints velocity holds

$$\dot{q} = J(q)^{\#}v + Z(q)\lambda, \tag{3}$$

Interested readers may find more details in [15, 18].

3   Control Design

The purpose of the control is to track a desired trajectory for the pose (position plus orientation) of the end-effector in the Cartesian space while fulfilling a compliant behaviour for the manipulator without interfering with the main task.

---

[2]Notice that, during the experiments, a damped least-squares solution is anyway employed.

The first level of the designed controller is a classic inverse dynamics [19]

$$\boldsymbol{\tau} = \boldsymbol{B}(\boldsymbol{q})\boldsymbol{u}_q + \boldsymbol{C}(\boldsymbol{q}, \dot{\boldsymbol{q}})\dot{\boldsymbol{q}} + \boldsymbol{g}(\boldsymbol{q}), \tag{4}$$

where $\boldsymbol{u}_q \in \mathbb{R}^n$ is a new virtual control input. Replacing (4) into (2) yields the following closed-loop dynamics

$$\ddot{\boldsymbol{q}} = \boldsymbol{u}_q - \boldsymbol{B}(\boldsymbol{q})^{-1}\boldsymbol{\tau}_{ext}. \tag{5}$$

The following command acceleration can be thus designed [18]

$$\boldsymbol{u}_q = \boldsymbol{J}(\boldsymbol{q})^{\#}\left(\boldsymbol{u}_v - \dot{\boldsymbol{J}}(\boldsymbol{q})\dot{\boldsymbol{q}}\right) + \boldsymbol{Z}(\boldsymbol{q})\left(\boldsymbol{u}_\lambda - \dot{\boldsymbol{Z}}(\boldsymbol{q})^{\#}\dot{\boldsymbol{q}}\right), \tag{6}$$

where $\boldsymbol{u}_v \in \mathbb{R}^6$ and $\boldsymbol{u}_\lambda \in \mathbb{R}^r$ are new virtual control inputs in the Cartesian and in the null-space, respectively. The closed-loop dynamics can be projected in both spaces by substituting (6) into (5). Afterwards, multiplying both sides of the resulting equation by $\boldsymbol{J}(\boldsymbol{q})$ and $\boldsymbol{Z}(\boldsymbol{q})^{\#}$, respectively, yields

$$\dot{\boldsymbol{v}} = \boldsymbol{u}_v - \boldsymbol{J}(\boldsymbol{q})\boldsymbol{B}(\boldsymbol{q})^{-1}\boldsymbol{\tau}_{ext}, \tag{7}$$

$$\dot{\boldsymbol{\lambda}} = \boldsymbol{u}_\lambda - \boldsymbol{Z}(\boldsymbol{q})^{\#}\boldsymbol{B}(\boldsymbol{q})^{-1}\boldsymbol{\tau}_{ext}. \tag{8}$$

The design of $\boldsymbol{u}_v$ and $\boldsymbol{u}_\lambda$ is addressed in the following.

3.1 Design of $\boldsymbol{u}_v$

Let $\widetilde{\boldsymbol{p}} = \boldsymbol{p}_d - \boldsymbol{p}_e \in \mathbb{R}^3$ and $\dot{\widetilde{\boldsymbol{p}}} = \dot{\boldsymbol{p}}_d - \dot{\boldsymbol{p}}_e \in \mathbb{R}^3$ be the position and the linear velocity error vectors, respectively. Moreover, set $\dot{\boldsymbol{v}}_d = \left[\ddot{\boldsymbol{p}}_d^{\mathrm{T}} \ \dot{\boldsymbol{\omega}}_d^{\mathrm{T}}\right]^{\mathrm{T}} \in \mathbb{R}^6, \boldsymbol{e}_v = \left[\dot{\widetilde{\boldsymbol{p}}}^{\mathrm{T}} \ \widetilde{\boldsymbol{\omega}}^{\mathrm{T}}\right]^{\mathrm{T}} \in \mathbb{R}^6$, and $\boldsymbol{e}_t = \left[\widetilde{\boldsymbol{p}}^{\mathrm{T}} \ \widetilde{\boldsymbol{o}}^{\mathrm{T}}\right]^{\mathrm{T}} \in \mathbb{R}^6$. Then, the input term $\boldsymbol{u}_v$ in the Cartesian space can be designed as

$$\boldsymbol{u}_v = \dot{\boldsymbol{v}}_d + \boldsymbol{D}_v \boldsymbol{e}_v + \boldsymbol{K}_v \boldsymbol{e}_t - \boldsymbol{J}(\boldsymbol{q})\boldsymbol{B}(\boldsymbol{q})^{-1}\boldsymbol{\gamma}, \tag{9}$$

with $\boldsymbol{K}_v = \mathrm{diag}(\boldsymbol{K}_p, \boldsymbol{K}_o) \in \mathbb{R}^{6 \times 6}$, where $\boldsymbol{K}_p \in \mathbb{R}^{3 \times 3}$ is a positive definite diagonal gain matrix and $\boldsymbol{K}_o \in \mathbb{R}^{3 \times 3}$ is an invertible matrix, $\boldsymbol{D}_v \in \mathbb{R}^{6 \times 6}$ a positive definite diagonal gain matrix, and the vector $\boldsymbol{\gamma} \in \mathbb{R}^n$ is defined such that its time derivative is equal to

$$\dot{\boldsymbol{\gamma}} = -\boldsymbol{K}_I(\boldsymbol{\gamma} + \boldsymbol{\tau}_{ext}) - \boldsymbol{K}_\gamma^{-1}\boldsymbol{B}(\boldsymbol{q})^{-1}\boldsymbol{J}(\boldsymbol{q})^{\mathrm{T}}\boldsymbol{e}_v, \tag{10}$$

where $\boldsymbol{K}_I \in \mathbb{R}^{n \times n}$ and $\boldsymbol{K}_\gamma \in \mathbb{R}^{n \times n}$ are positive definite diagonal gain matrices.

Examining (10) it is possible to notice that a measurement of $\boldsymbol{\tau}_{ext}$ is needed. Nonetheless, it is possible to show that (10) has a closed-form solution

$$\gamma(t) = K_I \left(B(q)\dot{q} - \int_0^t \left(\tau + C(q,\dot{q})^{\mathrm{T}}\dot{q} - g(q) + \gamma(\sigma) \right) d\sigma \right)$$

$$- K_\gamma^{-1} \int_0^t B(q)^{-1} J(q)^{\mathrm{T}} e_v d\sigma,$$

(11)

that can be directly replaced in (9). Notice that the measurements of neither $\ddot{q}$ nor τ_{ext} are required in (11). The first part of (11) is equal to the momentum-based observer introduced in [3]. The last part has been instead added to cope with the employed singularity-free orientation representations.

By substituting the designed control law (9) into (7), it is straightforward to write down the corresponding Cartesian space closed-loop equation as

$$\dot{e}_v + D_v e_v + K_v e_t = J(q)B(q)^{-1}e_\gamma,$$

(12)

where $e_\gamma = \gamma + \tau_{ext} \in \mathbb{R}^n$.

3.2 Design of u_λ

Let $\lambda_d \in \mathbb{R}^r$ and $\dot{\lambda}_d \in \mathbb{R}^r$ be the null-space desired velocity and acceleration vectors, respectively. Define with $q_d \in \mathbb{R}^n$ the time-varying desired value of the joint positions. This should be planned accordingly to (p_d, R_d) through the robot inverse kinematics. Moreover, let $e_\lambda = \lambda_d - \lambda \in \mathbb{R}^r$ and $e_q = q_d - q \in \mathbb{R}^n$ be the null-space velocity and the joint configuration errors, respectively. As highlighted in [4], λ is not integrable and it is not thus possible to define a position error in the null-space. Therefore, the design of u_λ follows [16, 18] as

$$u_\lambda = \dot{\lambda}_d + \Lambda_\lambda(q)^{-1} \left((\mu_\lambda(q,\dot{q}) + D_\lambda) e_\lambda + Z(q)^{\mathrm{T}} (K_q e_q + D_q \dot{e}_q) \right),$$

(13)

where $K_q \in \mathbb{R}^{n \times n}$ and $D_q \in \mathbb{R}^{n \times n}$ are definite positive gain matrices, $\Lambda_\lambda(q) = Z(q)^{\mathrm{T}} B(q) Z(q) \in \mathbb{R}^{r \times r}$, and $\mu_\lambda(q,\dot{q}) = \left(Z(q)^{\mathrm{T}} C(q,\dot{q}) - \Lambda_\lambda(q)\dot{Z}(q)^{\#} \right) Z(q) \in \mathbb{R}^{r \times r}$. Notice that $\Lambda_\lambda(q)$ is positive definite, while $\dot{\Lambda}_\lambda(q) - 2\mu_\lambda(q,\dot{q})$ is skew-symmetric [18]. By substituting (13) into (8), it is possible to write down the null-space closed-loop equation as

$$\Lambda_\lambda(q)\dot{e}_\lambda + (\mu_\lambda(q,\dot{q}) + D_\lambda) e_\lambda + Z(q)^{\mathrm{T}} (K_q e_q + D_q \dot{e}_q) = Z(q)^{\mathrm{T}} \tau_{ext}.$$

(14)

This acts as an impedance controller against the projection τ_{ext} in the null-space. The matrix $\Lambda_\lambda(q)$ represents the inertia in the null-space, while $\mu_\lambda(q,\dot{q})$ the Coriolis matrix in the same space. These matrices cannot be modified, as instead D_λ, K_q and D_q that can be tuned to specify the desired null-space behaviour.

4   Proof of Stability

Recalling the designed control inputs (9), (11) and (13), and the resulting closed-loop system equations (12) and (14), to rigorously prove the stability of the system, the state $x = (e_q, e_t, e_v, e_\gamma, e_\lambda) \in \mathbb{R}^m$, with $m = 2n + r + 12$, has to asymptotically go to zero. Conditional stability [16] is the concept employed in the provided proof. Therefore, it is worth introducing the following theorem.

Theorem 1 *Let $\bar{x} = \mathbf{0}_m$ be an equilibrium point of the system $\dot{x} = f(x)$, with $f(x) \in \mathbb{R}^m$. Then, $\bar{x}$ is asymptotically stable if, in a neighbourhood Ω of $\bar{x}$, there exists a function $V \in \mathcal{C}^1$ such that*

1. *$V(x) \geq 0$ for all $x \in \Omega$ and $V(\bar{x}) = 0$;*
2. *$\dot{V}(x) \leq 0$ for all $x \in \Omega$;*
3. *on the largest positive invariant set $\mathcal{L} \subseteq \mathcal{Y} = \{x \in \Omega : \dot{V}(x) = 0\}$, the system is asymptotically stable.*

Proof See [11].   □

The following proposition proves the stability of (12) and (14).

Proposition 1 *Let K_v be a block-diagonal invertible matrix, while let D_v, K_I, K_γ, K_q, D_q, D_λ be diagonal and positive definite matrices. Assume that $\dot{\tau}_{ext} = \mathbf{0}_n$ and $\lambda_d = Z(q)^\#\dot{q}_d$. Then, considering a redundant robot arm whose dynamic model is given by (2), the control laws (4), (9), (11) and (13) are able to*

1. *bring the state $x = (e_q, e_t, e_v, e_\gamma, e_\lambda)$ asymptotically to zero if $\tau_{ext} = \mathbf{0}_n$, and q_d is chosen as fitting the pose (p_d, R_d) at each instant of time;*
2. *bring the state asymptotically to zero except e_q if $\tau_{ext} \neq \mathbf{0}_n$ and/or q_d is chosen such that the manipulator end-effector is not in the pose (p_d, R_d) at each instant of time. In this case q tends to $q^\star \neq q_d$, that belongs to the set of solutions locally minimizing the quadratic function $\| K_q e_q + D_q \dot{e}_q - \tau_{ext} \|^2$.*

Proof The proof is based on Theorem 1. Define the following scalar function in a neighbourhood Ω of the origin $\bar{x} = \mathbf{0}_m$

$$V(e_v, e_t, e_\gamma) = \frac{1}{2} e_v^\mathsf{T} e_v + \frac{1}{2} e_t^\mathsf{T} K_V e_t + \frac{1}{2} e_\gamma^\mathsf{T} K_\gamma e_\gamma + f_V(\alpha), \tag{15}$$

where $K_V = \mathrm{diag}(K_p, K_{V,2}) \in \mathbb{R}^{6\times 6}$, with $K_{V,2} \in \mathbb{R}^{3\times 3}$ a positive definite diagonal matrix, and $f_V(\alpha) \in \mathbb{R} \geq 0$. This last function depends on the chosen orientation representation. In the case of interest with $\tilde{o} = \tilde{\epsilon}$, then $f_V = k_\epsilon(\tilde{\eta} - 1)^2$, with $k_\epsilon > 0$. Notice that $V(\mathbf{0}_m) = 0$ and since $V(e_v, e_t, e_\gamma)$ is not defined on all the state variables, it is only semi-definite in Ω. This satisfies the first point of the Theorem 1. It is useful to compute the time derivative of e_t that is equal to

$$\dot{e}_t = \begin{bmatrix} \dot{\tilde{p}}^\mathsf{T} & f_m(\alpha, \tilde{\omega})^\mathsf{T} \end{bmatrix}^\mathsf{T}, \tag{16}$$

with $f_m(\alpha, \widetilde{\omega})$ chosen on the base of the available representation for the orientation. Having in mind (1), it yields $f_m(\alpha, \widetilde{\omega}) = L_q(\alpha)\widetilde{\omega}$. Moreover the assumption $\dot{\tau}_{ext} = 0_n$ yields

$$\dot{e}_\gamma = \dot{\gamma}. \tag{17}$$

Therefore, choosing $K_o = k_\epsilon I_3$ and $K_{V,2} = 2k_\epsilon I_3$, taking into account (12), (16) and (17), deriving (15) with respect to time yields $\dot{V} = -e_v^T D_v e_v - e_\gamma^T K_\gamma K_I e_\gamma$, which is negative semi-definite in Ω, satisfying the second point of Theorem 1.

Now the objective is to fulfil the third point of Theorem 1. Define the set $\mathcal{Y} = \{x \in \Omega : e_q, e_t, e_v = 0_6, e_\gamma = 0_r, e_\lambda\}$. Besides, define

$$V_{\mathcal{Y}} = \frac{1}{2}e_\lambda^T \Lambda_\lambda(q)e_\lambda + \frac{1}{2}e_t^T e_t + \frac{1}{2}e_q^T K_q e_q, \tag{18}$$

which it positive definite in $\mathcal{Y}$. In this set, since $e_v = 0_6$ then $v_d = v$. Moreover, having in mind (12) and since $e_\gamma = 0_r$ as well, this implies that $e_t = 0_6$. Because $\dot{e}_q = Ze_\lambda$ and considering (3), (14) and (16), the time derivative of (18) within $\mathcal{Y}$ is $\dot{V}_{\mathcal{Y}} = -e_\lambda^T \left(D_\lambda + Z^T D_q Z\right) e_\lambda + e_\lambda^T Z(q)^T \tau_{ext}$.

Initially, consider the first point of Proposition 1. Since $\tau_{ext} = 0_n$ and q_d is chosen such that the manipulator end-effector is located at (p_d, R_d) at each instant of time, then $\dot{V}_{\mathcal{Y}} < -\lambda^T D_\lambda \lambda$, which is negative semi-definite in $\mathcal{Y}$. Invoking the La Salles invariance principle and having in mind (14), it is straightforward to prove that $e_q \to 0_n$. Hence, for the above considerations, the system is asymptotically stable on the largest invariant set $\mathcal{L} \subseteq \mathcal{Y}$. This satisfies the third and last point of Theorem 1, and thus proves the first point of Proposition 1.

Finally, consider the second point of Proposition 1. Since $\tau_{ext} \neq 0_n$ and/or q_d is not chosen such that the end-effector is located at the desired configuration (p_d, R_d) at each instant of time, all the states variables of x goes asymptotically to zero except e_q. As a matter of fact $q \to q^\star$ as shown in [18]. This happens because the robot reaches a joints configuration $q^\star$ compatible with the main task (p_d, R_d) since $e_t = 0_6$ and minimizing the elastic potential energy $\|K_q e_q + D_q \dot{e}_q - \tau_{ext}\|^2$. The solution of such a minimization is found by solving $Z(q)^T \left(K_q e_q + D_q \dot{e}_q - \tau_{ext}\right) = 0_r$. Subsequently, $\dot{V}_{\mathcal{Y}}$ is less or equal to $-\lambda^T D_\lambda \lambda$, and thus e_λ tends to zero: this proves the second point of Proposition 1. $\square$

5   Experiments

Experimental validation has been carried out on a KUKA LWR4 with $n = 7$. For each case study described below the controller sample time is equal to 2 ms, while the gains have been experimentally tuned to the following values: $K_p = 150I_2$, $K_o = 150I_3$, $D_v = \text{blockdiag}\{20I_2, 10I_3\}$, $K_I = 10I_7$, $K_\gamma = 100I_7$, $K_q = 5I_7$, $D_\lambda = 5I_5$. The computation of $Z(q)$ has been carried out in a symbolic way thanks to the Mathematica software. Without loss of generality, only the unit quaternion representation is employed in the following.

Three case studies are investigated. The considered scenario is depicted in Fig. 1. In the first case study the robot end-effector has to follow the path from A to B, whose length is about 0.6 m, keeping fixed the intial orientation. The timing law along the path is a trapezoidal acceleration with the cruise acceleration set to 0.15 m/s^2 [19]. By following the path, the tool attached to the end-effector crashes against the obstacles. However, since only the two planar positions variables in Σ_i and the orientation are constrained (the vertical position is left free), the dimension of λ is $r = 2$. This allows a human operator to push the robot to avoid obstacles by changing the configuration in the null space. It is worth pointing out that, due to the human presence, it is not possible to guarantee a constant τ_{ext} during the experiments. Even if the developed theory shows asymptotic stability only for $\dot{\tau}_{ext} = \mathbf{0}_n$, the overall performance remains good as admirable in the next subsections. The desired $\boldsymbol{q}_d$ is chosen off-line such that at each instant of time the end-effector is kept on the designed path.

In the second case study, the robot performs the same task, but $\boldsymbol{K}_q$ is set to zero. Hence, the robot does not come back to the planned $\boldsymbol{q}_d$ when pushed.

In the third case study, the two obstacles at the boundaries are removed: only the central obstacle depicted in Fig. 1 is kept. The matrix $\boldsymbol{K}_q$ is set back to its tuned value. The same position path is followed but the orientation has to follow a desired trajectory as well. The initial and final quaternions are $\boldsymbol{\alpha}_{init} = \begin{bmatrix} 0.007 & 0.9336 & 0.3581 & -0.0007 \end{bmatrix}^T$ and $\boldsymbol{\alpha}_{fin} = \begin{bmatrix} 0.7071 & 0.7071 & 0 & 0 \end{bmatrix}^T$. An angle-axis orientation planner is then built [19], with a trapezoidal acceleration whose cruise value is 0.15 rad/s^2. Again, the planned movement leads to a collision between the tool attached at the end-effector and the central obstacle. A human operator can reconfigure the robot internally by pushing it to avoid the object since the vertical position is not constrained. Again, the dimension of λ is $r = 2$.

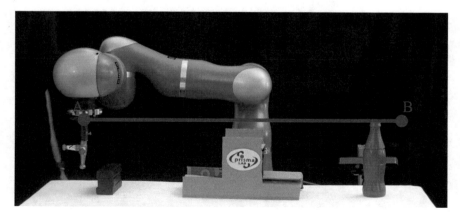

Fig. 1 The KUKA LWR4 employed for the experiments in the initial, and desired, configuration for the joints, and with the desired orientation for the end-effector

5.1 Case Study 1

Time histories related to this first case study are depicted in Fig. 2. The two planar
components of $\widetilde{\boldsymbol{p}}$ in Σ_i are shown in Fig. 2a: the overall graph error is between
± 0.02 m, while its maximum is reached at about 15 s when there is a relevant
interaction with the human operator, as it is also possible to check from Fig. 2d
where the γ term, resembling τ_{ext}, is represented. From Fig. 2d, it is also possible
to see that human interaction happens at around 1, 6, 10, 15 and 18 s to both avoid
obstacles and test robustness.

The orientation error $\widetilde{\boldsymbol{o}}$ is depicted in Fig. 2b through its geodesic measure o_g in
$SO(3)$ [10]. While the non-dimensional term $\widetilde{\boldsymbol{o}}$ is directly used in the control and
it is not a clearly understandable for a reader, the geodesic measure is expressed in
radians (or degree) and represents the distance of the deviation matrix $\widetilde{\boldsymbol{R}}$ from the
identity. Expressing with $\| \cdot \|_F$ the Frobenius norm, the geodesic measure can be
computed as $o_g = 1/\sqrt{2}\|\log \widetilde{\boldsymbol{R}}\|_F$. Within Sect. 2.2, and the references therein, it is
possible to find the relations linking $\widetilde{\boldsymbol{R}}$ with $\widetilde{\boldsymbol{o}}$. From the time history related to the

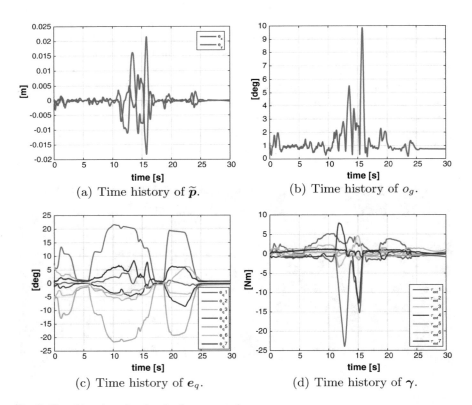

(a) Time history of $\widetilde{\boldsymbol{p}}$.

(b) Time history of o_g.

(c) Time history of $\boldsymbol{e}_q$.

(d) Time history of $\boldsymbol{\gamma}$.

Fig. 2 Time histories related to the first case study

first case study, it is remarkable that the orientation error is under the 3 deg for most of time. The maximum error is reached again around 15 s. It is worth highlighting that the error does not come back exactly to zero due to the presence of non-negligible joint friction and other small uncertainties that the system is not able to recover.

Finally, the time history of e_q is depicted in Fig. 2c. It is possible to appreciate that once the interaction with the human is ended, the manipulator tends to come back to the (time-varying) desired q_d.

5.2   Case Study 2

The time histories related to this case study are depicted in Fig. 3. The position error $\widetilde{p}$ is shown in Fig. 3a for the two planar components expressed in Σ_i: the overall graph error is between ± 0.015 m. Figure 3d shows instead the time history of the γ term, in which it is possible to see that human interaction happens at around 2 s, 8 s, 11 s, 13 s, 17 s, 22 s and 26 s to both avoid obstacles and test robustness. The geodesic measure

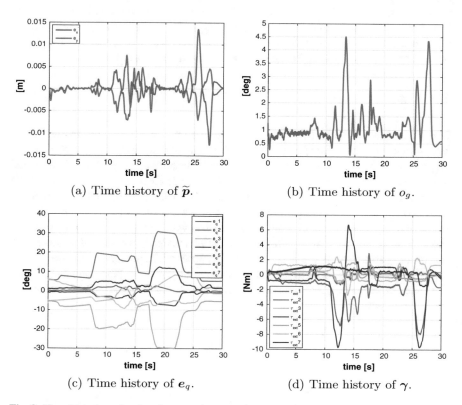

(a) Time history of $\widetilde{p}$.

(b) Time history of o_g.

(c) Time history of e_q.

(d) Time history of γ.

Fig. 3 Time histories related to the second case study

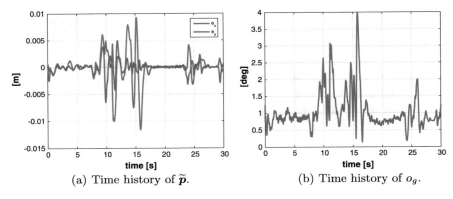

(a) Time history of $\widetilde{\boldsymbol{p}}$.   (b) Time history of o_g.

Fig. 4 Time histories related to the third case study

o_g related to $\widetilde{\boldsymbol{o}}$ is depicted in Fig. 2b. Such orientation error is lower than 2 deg for most of experiment. The maximum error is reached around 13 s and 26 s. Finally, the time history of $\boldsymbol{e}_q$ is depicted in Fig. 3c. It is possible to appreciate that once the interaction with the human is ended, the manipulator stays in the reached internal configuration since the gain $\boldsymbol{K}_q$ has been set to zero. The manipulator behaviour is pretty similar to the one achievable through a direct zero-force control law.

5.3 Case Study 3

Time histories related to this case study are depicted in Fig. 4. The position error $\widetilde{\boldsymbol{p}}$ is shown in Fig. 4a for the two planar components expressed in Σ_i: the overall graph error is between ± 0.01 m. The geodesic measure o_g related to $\widetilde{\boldsymbol{o}}$ is depicted in Fig. 4b. Such orientation error is lower than 3 deg for most of experiment. The maximum error of about 4 deg is reached around 16 s. The time histories for the γ term and $\boldsymbol{e}_q$ are not reported for brevity since they are very similar to the first case study, in which the end-effector desired behaviour was to track a path for the position and keep fixed the orientation. This last case study, instead, shows how the proposed controller is able to track the desired Cartesian pose, while preserving the possibility to act in the null-space with a compliance behaviour preserving the main task. These and other experiments, like pouring a water in a moving glass, are available in the related video-clip.[3]

---

[3]https://www.youtube.com/watch?v=PirdFEAE_D8.

6 Conclusion

In this paper, a framework to control a redundant manipulator in the Cartesian space for a tracking task is designed addressing a singularity-free representation for the orientation, and allowing the possibility to change the null-space configuration of the manipulator without affecting the main task. The designed controller does not need any exteroceptive sensors to accomplish the task, as well as no joints torque sensors are requested. Theory and experimental results bolster the effectiveness of the proposed control scheme.

Acknowledgements The research leading to these results has been supported by the RoDy-Man project, which has received funding from the European Research Council FP7 Ideas under Advanced Grant agreement number 320992. The authors are solely responsible for the content of this manuscript.

References

1. Bicchi, A., Tonietti, G.: Fast and soft arm tactics: dealing with the safety performance trade-off in robot arms design and control. IEEE Robot. Autom. Mag. **11**(2), 22–33 (2004)
2. Caccavale, F., Natale, C., Siciliano, B., Villani, L.: Resolved-acceleration control of robot manipulators: a critical review with experiments. Robotica **16**, 565–573 (1998)
3. de Luca, A., Albu-Schaffer, A., Haddadin, S., Hirzinger, G.: Collision detection and safe reaction with the DLR-III lightweight robot arm. In: 2006 IEEE/RSJ International Conference on Intelligent Robots Systems. pp. 1623–1630. Beijing, C (2006)
4. de Luca, A., Oriolo, G.: Nonholonomic behavior in redundant robots under kinematic control. IEEE Trans. Robot. Autom. **13**(5), 776–782 (1997)
5. De Santis, A., Siciliano, B., De Luca, A., Bicchi, A.: An atlas of physical human-robot interaction. Mechan. Mach. Theory **43**, 2008 (2008)
6. Feil-Seifer, D., Mataric, J.: Defining socially assistive robotics. In: 9th International Conference on Rehabilitation Robotics. pp. 465–468. Chicago, IL, USA (2005)
7. Ficuciello, F., Villani, L., Siciliano, B.: Variable impedance control of redundant manipulators for intuitive human-robot physical interaction. IEEE Trans. Robot. **31**(4), 850–863 (2015)
8. Haddadin, S., aLBU Schäffer, A., Hirzinger, G.: Requirements for safe robots: measurements, analysis and new insights. Int. J. Robot. Res. **28**, 1507–1527 (2009)
9. Hogan, N.: Impedance control: an approach to manipulation: Parts I–III. ASME J. Dyn. Syst. Measur. Control **107**(2), 1–24 (1985)
10. Huynh, D.: Metrics for 3D rotations: comparison and analysis. J. Math. Imaging Vis. **35**, 155–164 (2009)
11. Iggidr, A., Sallet, G.: On the stability of nonautonomous systems. Automatica **39**, 167–171 (2003)
12. Khatib, O.: Inertial properties in robotic manipulation: an object-level framework. Int. J. Robot. Res. **14**(1), 19–36 (1995)
13. Maaroof, O.W., Dede, M.: Physical human-robot interaction: increasing safety by robot arms posture optimization. In: Parenti-Castelli, V., Schiehlen, W. (eds.) ROMANSY 21-Robot Design, Dynamics and Control. CISM International Centre for Mechanical Sciences (Courses and Lectures), pp. 329–337. vol 569. Springer, Cham (2016)
14. Oh, Y., Chung, W., Youm, Y.: Extended impedance control of redundant manipulators based on weighted decomposition of joint space. J. Intell. Robot. Syst. **15**(5), 231–258 (1998)

15. Ott, C.: Control of nonprehensile manipulation. In: Cartesian impedance control of redundant and flexible joint robots. Springer Tracts in Advanced Robotics, vol. 49. Springer, New York, NY, USA (2008)
16. Ott, C., Kugi, A., Nakamura, Y.: Resolving the problem of non-integrability of null-space velocities for compliance control of redundant manipulators by using semi-definite lyapunov functions. In: 2008 IEEE International Conference on Robotics and Automation, pp. 1999–2004. Pasadena, CA, USA (2008)
17. Sadeghian, H., Villani, L., Keshmiri, M., Siciliano, B.: Multi-priority control in redundant robotic systems. In: 2011 IEEE/RSJ International Conference on Intelligent Robots and Systems, pp. 3752–3757. San Francisco, CA, USA (2011)
18. Sadeghian, H., Villani, L., Keshmiri, M., Siciliano, B.: Task-space control of robot manipulators with null-space compliance. IEEE Trans. Robot. **30**(2), 493–506 (2014)
19. Siciliano, B., Sciavicco, L., Villani, L., Oriolo, G.: Robotics: Modelling Planning and Control. Springer, London, UK (2009)
20. Vigoriti, F., Ruggiero, F., Lippiello, V., Villani, L.: Control of redundant arms with null-space compliance and singularity-free orientation representation. Robot. Auton. Syst. **100**, 186–193 (2018)

Repetitive Jumping Control for Biped Robots via Force Distribution and Energy Regulation

Gianluca Garofalo and Christian Ott

Abstract The paper presents a new control law to initiate and stop a sequence of repetitive jumps in elastically actuated legged robots. The control approach relies on three control strategies, which are effectively combined to realize the task: hierarchical task space decomposition, balancing force redistribution and energy regulation. A toy example motivates the interconnection of the three parts, while an experimental evaluation is used to corroborate the effectiveness of the controller on a complex biped robot.

Keywords Humanoid robots · Task hierarchy control · Balancing control
Energy control · Jumping robots

1 Introduction

Although jumping robots are present on the robotic scene since a while, during the years this task has been mainly achieved by quadrupedal robots for which the balancing problem is notably simplified. Both bipedal and quadrupedal robots usually use simplified models or offline optimization in order to design force profiles and reference trajectories [14, 16]. Few of them include elastic elements in the legs [9], but it is not clear how the controller is actively using the springs beside the benefit from improved shock absorption at the impacts. This is one of the points on which this paper will focus.

The notion of elastic joint robots has a long history in robotics. Nevertheless, while in the seminal publications by Spong [18] and De Luca [2], the joint elasticity

G. Garofalo (✉) · C. Ott
German Aerospace Center (DLR), Institute of Robotics and Mechatronics,
Münchner Strasse 20, 82234 Wessling, Germany
e-mail: gianluca.garofalo@dlr.de
URL: http://www.dlr.de/rmc/rm

C. Ott
e-mail: christian.ott@dlr.de

© Springer International Publishing AG, part of Springer Nature 2019
F. Ficuciello et al. (eds.), *Human Friendly Robotics*, Springer Proceedings
in Advanced Robotics 7, https://doi.org/10.1007/978-3-319-89327-3_3

was originally treated as a disturbance of the rigid-body dynamics, more recent drive concepts (like series elastic actuators [15] or variable impedance actuators [19]) deliberately introduce elasticity for implementing torque control, increasing physical robustness, or reaching high output velocities. Therefore, legged robots have been both the application and motivation for the design of such innovative actuation systems. Elastically actuated legged robots are underactuated systems and, therefore, they are challenging to control. The underactuation is a direct consequence of both their floating base nature and the presence of the springs in the joints.

In this paper, the complete underactuated problem is considered. The robot is neither fixed to the floor, nor rigidly actuated. A bipedal robot is used as case study (see Fig. 1). The derivation of the control law is conceptually split in two steps. In the first one, a hierarchical task space decomposition [12] and an energy regulation [6] are used to generate a first torque reference. In the second step, a balancing force redistribution [10] modifies the torque reference to ensure the feasibility of the controller. The simplifying hypothesis used throughout the paper requires that the balancing torques can be computed assuming to have a rigid joint, i.e. the current spring torques cannot make the robot lose balance, while the flexibility is considered for the rest of the derivation. The assumption is reasonable since the spring torque is part of the state and cannot be instantaneously changed by control.

The main contribution of the paper is a control law that directly uses the elasticity in the joints in order to produce a controlled sequence of jumps, while considering the whole dynamic model of the robot. The resulting jumps are not pre-planned, but

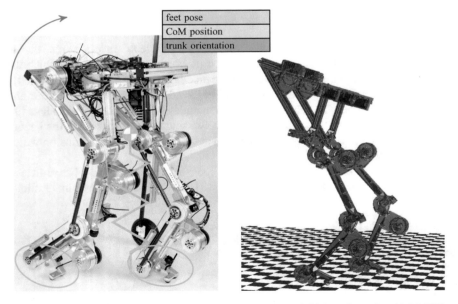

Fig. 1 C-Runner: real test-bed (left) and simulated model in the initial configuration (right) [11]

a dynamical result of the actions of the springs, whose torques are shaped trough the control action, as it is presented in Sect. 6. As the method relies on the interconnection of three control approaches, each one of them is shortly reviewed in Sects. 2–4, while Sect. 5 serves as motivation on how to interconnect these components. The paper is concluded with a final discussion and outline of future work in Sect. 7.

1.1 Notation and Model

With a slight abuse of notation, throughout the paper it will be used M to denote the symmetric and positive definite inertia matrix, C a Coriolis matrix satisfying $\dot{M} = C + C^T$ and g the gravity torque vector, with suitable dimensions depending on the type of model considered (i.e. fixed or floating base). When dealing with elastic joints, these matrices have to be understood as link-side quantities, with B denoting instead the inertia of the motors. The torques τ_m produced by the motors are an input to the system. These are directly the torques τ applied to the link for a rigid joint robot (i.e. $\tau = \tau_m$), while for elastic joints the latter are the torques produced by the springs. In addition to the motor torques, the external wrenches stacked in w_f and mapped through J_f^T complete the inputs to the system. The most general case considered in this paper is a *floating base robot with linear elastic joints*:

$$M(x)\dot{v} + C(x, v)v + g(x) = Q^T\tau + J_f^T(x)w_f , \tag{1a}$$

$$B\ddot{\theta} + \tau = \tau_m , \tag{1b}$$

$$\tau = K\left(\theta - q\right) , \tag{1c}$$

where $\theta, x, \dot{\theta}, v$ constitute together the state of the system, being θ the motor position and x the link-side configuration, which includes the link position q and the floating base coordinates. The matrix Q selects the joint velocities $\dot{q}$ out of all the velocity coordinates v, i.e. $\dot{q} = Qv$. Finally, $K = \text{diag}(K_i)$ is the joint stiffness matrix, with K_i the stiffness constant of the i-th joint. To the assumptions for the derivation of the reduced elastic joint model in [18], the following one is added [6]

Assumption 1

$$\left\| \frac{\partial g}{\partial q} \right\| < \min_i K_i \qquad \forall x \in \mathbb{X}$$

where the subset $\mathbb{X}$ of the state space in which all the prismatic joints are kept bounded [1] will always be considered throughout the whole paper.

2   The Task Hierarchy Controller

The task hierarchy controller presented in [12] extends the results in [13] in order
to cope with the different priorities that the tasks of a *fixed base manipulator with
rigid joints* can have. Given the joint coordinates $q \in \mathbb{R}^n$, let $y(q)$ indicate the whole
task coordinates, so that $y(q) \in \mathbb{R}^n$, and let $y_i(q)$ be one of the r subtasks. Given
the mapping

$$\dot{y}_i = J_i(q)\dot{q} \quad 1 \le i \le r, \tag{2}$$

assume additionally that each of the Jacobian matrices $J_i(q) \in \mathbb{R}^{m_i \times n}$ is full rank, as
well as each Jacobian matrix obtained stacking any $J_i(q)$. A coordinate transforma-
tion can then be considered, which replaces the joint velocities $\dot{q} \in \mathbb{R}^n$ with $\zeta \in \mathbb{R}^n$
obtained through the extended Jacobian matrix $\bar{J}_N(q) \in \mathbb{R}^{n \times n}$, as

$$\zeta = \begin{bmatrix} \zeta_1 \\ \vdots \\ \zeta_r \end{bmatrix} = \begin{bmatrix} \bar{J}_1(q) \\ \vdots \\ \bar{J}_r(q) \end{bmatrix} \dot{q} = \bar{J}_N(q)\dot{q}, \tag{3}$$

where $\bar{J}_1 = J_1$ and $\zeta_1 = \dot{y}_1$, while the others are nullspace velocities $\zeta_i \in \mathbb{R}^{m_i}$. The
latter are designed such that they represent the original task as closely as possible,
but they are inertially decoupled from the tasks with a higher priority level [12]. The
extended Jacobian matrix is invertible, so that

$$\dot{q} = \bar{J}_N^{-1}(q)\zeta = \sum_{i=1}^{r} Z_i^T(q)\zeta_i, \tag{4}$$

where $Z_1^T(q) = J_1^{+M}(q) := M^{-1}J_1^T\left(J_1M^{-1}J_1^T\right)^{-1}$, while $Z_i(q) = J_i(q)M^{-1}$
$\Upsilon_i(q)$ for $2 \le i \le r$ and $\Upsilon_i(q)$ the classical nullspace projector [17].
    Using this coordinate transformation the system can be written as

$$\Lambda(q)\dot{\zeta} + \Gamma(q,\dot{q})\zeta = \bar{J}_N^{-T}(q)\left(\tau - g(q)\right), \tag{5}$$

where the matrices $\Lambda(q)$ and $\Gamma(q,\dot{q})$ are block diagonal and with r blocks (one for
each subtask) on the main diagonal. The same stability property as in [13] holds for
the closed loop system obtained through the feedback control law in [12], which can
easily be rearranged as

$$\tau = \bar{J}_N^T \left(\bar{\Gamma} \begin{bmatrix} \zeta_1 \\ \vdots \\ \zeta_r \end{bmatrix} + \begin{bmatrix} Z_1\left(g + J_1^T\phi_1\right) \\ \vdots \\ Z_r\left(g + J_r^T\phi_r\right) \end{bmatrix} \right), \tag{6}$$

where $\bar{\boldsymbol{\Gamma}}$ is obtained from $\boldsymbol{\Gamma}$ setting the blocks on the main diagonal to zero[1] and $\phi_i(\boldsymbol{y}_i, \dot{\boldsymbol{y}}_i)$ is a PD term regulating the task coordinates $\boldsymbol{y}_i$ to their desired value $\boldsymbol{y}_{d_i}$. In addition to [13], here the transient response is guaranteed to have no inertial coupling between the different tasks.

3 The Balancing Controller

In [10] a balancing controller based on the task hierarchy controller was proposed for a *humanoid robot with rigid joints*. The goal is to have the CoM of the robot and the hip orientation (CoM task) in a given configuration in space. The joint configuration should also be as close as possible to a desired one (posture task). The CoM task together with the posture task completely define the configuration of the floating base robot. Due to the restrictions imposed to the system by the contact state, the posture task might not be fully fulfilled. Assuming to have some of the end effectors of the robot in contact with the environment and some free to move in space, the requirements are to produce balancing wrenches with the first (balancing task) and an impedance behavior with the latter (interaction task). The balancing wrenches have to counteract the effects of all the other task and compensate the gravity, ensuring that the base does not need to be actuated. Indicating by ϕ_i the correspondent task force, the tasks are stacked in the task hierarchy as: balancing task (ϕ_1), interaction task (ϕ_2), CoM task (ϕ_3), posture task (ϕ_4), where all the ϕ_i are chosen as PD terms, except for ϕ_1. The latter is determined through an optimization problem, which guarantees the feasibility of the control law, i.e. it makes sure that each balancing end effector does not lift off, slip or tilt, as well as ensuring that no forces at the floating base are commanded. In fact, the main difference compared to the task hierarchy controller in [12] is that $\boldsymbol{u} := \boldsymbol{Q}^T \boldsymbol{\tau}$ cannot be fully chosen, since $\boldsymbol{u} = \begin{bmatrix} \boldsymbol{0}^T & \boldsymbol{\tau}^T \end{bmatrix}^T$.

The balancing controller in [10] is designed using a frame attached to the CoM and with orientation given by the hip frame, so that in this coordinates the gravity torque simplifies to $\begin{bmatrix} m\boldsymbol{g}_0^T & \boldsymbol{0}^T \end{bmatrix}^T$, i.e. the weight of the robot with m its total mass. Using a quasi-static argument,[2] the controller (6) becomes

$$\begin{bmatrix} \boldsymbol{0} \\ \boldsymbol{\tau} \end{bmatrix} = \begin{bmatrix} m\boldsymbol{g}_0 \\ \boldsymbol{0} \end{bmatrix} + \begin{bmatrix} \boldsymbol{\Xi}_u \\ \boldsymbol{\Xi}_a \end{bmatrix} \boldsymbol{\phi} \,, \tag{7}$$

where

$$\boldsymbol{\Xi} = \begin{bmatrix} \boldsymbol{\Xi}_u \\ \boldsymbol{\Xi}_a \end{bmatrix} = \begin{bmatrix} \bar{\boldsymbol{J}}_{N_1}^T \boldsymbol{Z}_1 \boldsymbol{J}_1^T & \cdots & \bar{\boldsymbol{J}}_{N_r}^T \boldsymbol{Z}_r \boldsymbol{J}_r^T \end{bmatrix} \,, \tag{8}$$

---

[1]i.e. it is a skew-symmetric matrix.

[2]i.e. neglecting the effects due to the Coriolis matrix.

and all the forces have been stacked in ϕ. The optimization problem can then be formulated as

$$\min_{\phi_1} \left(\phi_1 - \phi_{d_1}\right)^T W \left(\phi_1 - \phi_{d_1}\right) \tag{9a}$$

$$\text{s.t. } m g_0 + \Xi_u \phi = 0 \tag{9b}$$

$$f_{min,i} \leq f_{i,\perp} \tag{9c}$$

$$\left\| f_{i,\|} \right\| \leq \mu_i f_{i,\perp} \tag{9d}$$

$$CoP_i(f_i) \in \mathcal{S}_i \tag{9e}$$

$$\tau_{min} \leq \Xi_a \phi \leq \tau_{max} , \tag{9f}$$

where the cost function minimizes the deviation from a default wrench distribution[3] ϕ_{d_1} with W a block-diagonal (one block for each task) positive definite weighting matrix. The equality constraint is derived from the first equation in (7), while from the second it follows

$$\tau = \Xi_a \phi , \tag{10}$$

that is the control law applied to the robot. The last inequality constraint is, therefore, ensuring that the resulting joint torques stay within the limitations of the hardware. Finally, the other constraints take into account the contact model. There, $f_{i,\perp}$ and $f_{i,\|}$ denote the components of the contact force f_i perpendicular and parallel to the contact surface $\mathcal{S}_i$. The unilaterality of the contact is taken into account by limiting the normal component to $f_{min,i} \geq 0$. To prevent the end effector from slipping, $f_{i,\|}$ is limited via the friction coefficient μ_i, while tilting is avoided restricting the center of pressure (CoP) of each end effector to $\mathcal{S}_i$. Each f_i enters in the dynamic model as part of w_f.

4   The Energy Controller

In [6] we introduced a nonlinear dynamic state feedback to initiate and stop a periodic motion for an *elastically actuated manipulator*, therefore extending our previous results for rigid joints [5, 8] and for a single elastic joint [3]. This periodic motion is due to the presence of an asymptotically stable limit cycle generated via a nonlinear feedback of an energy function, while simultaneously forcing the system to evolve on a submanifold of the configuration space. By taking into account the physical total potential energy U (i.e. gravitational plus elastic), the controller in [6] makes use of the energy stored in the springs. In particular, the energy function is given by

---

[3]Typically, this value is chosen to be an equally distributed gravity compensation between all the end effector in contact with the environment.

$$H(\chi) := \frac{1}{2}\dot{q}^T M \dot{q} + U(\eta, q) - U(\eta, \bar{q}_y(\eta)), \tag{11}$$

where $\eta, \dot{\eta} \in \mathbb{R}^n$ are the state variables of the dynamic controller. The function $\bar{q}_y(\eta)$ is given by

$$\bar{q}_y(\eta) := \arg\min_q U(\eta, q)$$
$$\text{s.t. } y_1(q) = 0, \tag{12}$$

which is well defined thanks to Assumption 1 and satisfies $\dot{U}(\eta, \bar{q}_y(\eta)) = \dot{\eta}^T K \left(\eta - \bar{q}_y(\eta) \right)$, as shown in [6]. The function $y_1 : \mathbb{R}^n \to \mathbb{R}^{n-1}$ has full rank Jacobian matrix, so that $y_1(q) = 0$ defines a submanifold. Forcing the robot to this submanifold corresponds to reach the configuration in which the robot will perform the periodic oscillation. The dynamic equations of the controller are

$$B\ddot{\eta} + K_H \tilde{H} K\left(\bar{q}_y(\eta) - q\right) + D_\eta \dot{\eta} + K_\eta \tilde{\eta} = 0 \tag{13a}$$

$$\tau_m = \tau_d + K\left(\eta - q\right) - K_H \tilde{H} K\left(\bar{q}_y(\eta) - q\right) + B K^{-1}\left(\ddot{\tau}_d - D_\tau \dot{\tilde{\tau}} - K_\tau \tilde{\tau}\right) - D_\eta \dot{\eta} - K_\eta \tilde{\eta}, \tag{13b}$$

where the positive scalar K_H and the symmetric, positive definite matrices $K_\tau, D_\tau, K_\eta, D_\eta \in \mathbb{R}^{n \times n}$ are control gains and $\tau_m \in \mathbb{R}^n$ is the output (feedback to the robot as in Fig. 2). Additionally, $\tilde{H} = H - H_d$, $\tilde{\eta} = \eta - \eta_d$, with $H_d \in \mathbb{R}$ and $\eta_d \in \mathbb{R}^n$ constant desired values, while $\tilde{\tau} = K\left(\theta - \eta\right) - \tau_d$ being $\tau_d \in \mathbb{R}^n$ the input function. The latter is computed based on the state of the system and of the controller itself [6] and, loosely speaking, it is responsible to enforce the $n - 1$ virtual constraints given

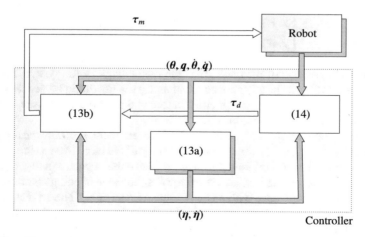

Fig. 2 Closed loop system

by $y_1(q) = 0$. Using the notation from Sect. 2, it is straightforward to rewrite τ_d as

$$\tau_d = \bar{J}_N^T \left(\bar{\Gamma}\zeta + \begin{bmatrix} Z_1\left(g - K(\eta - q) - J_1^T(K_1 y_1 + D_1 \dot{y}_1)\right) \\ 0 \end{bmatrix} \right), \qquad (14)$$

where K_1 and D_1 are the coefficients of the proportional and derivative action of the PD term. In [6] it is shown that the resulting closed-loop system has an asymptotically stable solution consisting of an equilibrium point, if $H_d = 0$, or a limit cycle, if $H_d > 0$.

5   A Toy Model for Jumping

Our previous works [3, 6, 8] guarantee the presence of an asymptotically stable limit cycle for the nominal system, i.e. without uncertainties. Although not formally proven in the stability analysis, the experiments therein show that periodicity is achieved even in case of oscillations around the desired value of the energy, due to unmodeled disturbances. On the other hand, in [7] we showed that a form of energy regulation combined with a control approach based on the Spring Loaded Inverted Pendulum (SLIP), was able to produce a periodic walking pattern for a simulated biped robot with impacts. These works motivate the idea behind the use of energy regulation for a jumping robot. This idea can be exemplified using a 1-dimensional SLIP model, i.e. the simplest scenario in which elasticity and energy regulation can be combined to obtain a jumping behavior. The system is represented by a single massless spring connected to a mass constrained to move along the vertical direction. The model of the system is:

$$\begin{cases} \ddot{y} + g = -K_h \tilde{H}\dot{y} - k\left(y - l_0\right) & y \le l_0 \\ \ddot{y} + g = 0 & y > l_0 \end{cases}, \qquad (15)$$

where y is the position of the mass, H is the total energy of the system, l_0 the rest length of the spring, k the stiffness constant and g the gravitational acceleration. In addition, it is assumed that the velocities before and after the impact are related by a certain coefficient of restitution α, i.e. $\dot{y}^+ = \alpha\dot{y}^-$.

Depending on the values of α and K_h, the different phase portraits are reported in Fig. 3. Firstly, in both cases, the strong asymmetry of the plot is due to the switching of the dynamics. The discontinuity, instead, is due to the impact. Finally, while with a high K_h (left) the system is able to reach the desired value of energy before the next take off, this is not the case with a small value of K_h (right). Nevertheless, a limit cycle is reached for which the energy lost at the impact balances the one injected before the take off. The energy for the two cases is plotted in Fig. 4.

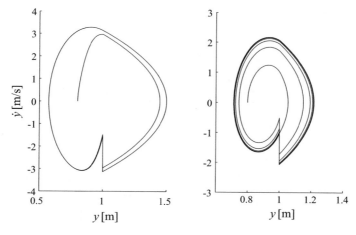

Fig. 3 Phase portrait of the toy model for $K_h = 5 \frac{s}{kgm^2}$ (left) and $K_h = 0.5 \frac{s}{kgm^2}$ (right)

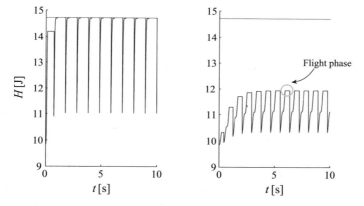

Fig. 4 Total energy and desired value for $K_h = 5 \frac{s}{kgm^2}$ (left) and $K_h = 0.5 \frac{s}{kgm^2}$ (right). In the right plot, the energy is constant only during the flight phase

Controlling the energy is therefore not only beneficial to exploit the presence of the elastic joints (as shown in [3, 6]), but it also provides a way to cope with an uncertain impact model. In this simple example, when the energy is quickly regulated to the desired value, the latter is also directly responsible for the jumping height. In Fig. 5, the actual and desired height are plotted, with the latter obtained from the desired energy value. For a given energy loss, the maximum reachable height is limited by the speed at which the energy can be reinjected in the system.

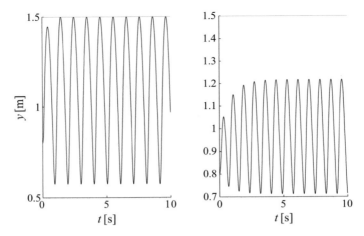

Fig. 5 Jumping height for $K_h = 5 \frac{s}{\text{kgm}^2}$ (left) and $K_h = 0.5 \frac{s}{\text{kgm}^2}$ (right)

6   The Jumping Controller

The main idea behind the jumping controller is to make the vertical motion of the robot's CoM behave similarly to the toy model of the previous section. The robot will be constrained in such a way that a vertical oscillation of the CoM is the only allowed motion. Given these constraints and the presence of the elasticity in the system, the model is conceptually similar to a 1-dimensional SLIP model, as it is depicted in Fig. 1 for the C-Runner. The regulation of the energy will then produce the desired oscillatory behavior of a jumping motion, as seen in Sect. 5. The input function of the energy controller will be chosen in the same fashion as in (6), in order to enforce the virtual constraints. As in Sect. 3, in the first stage the underactuation and the contact model will not be taken into account. An optimization problem will later on consider these restrictions by redistributing ϕ to generate feasible contact forces.

Starting from the dynamic equations 1 of the model, the dynamic state feedback controller from Sect. 4 is used. This leads to the closed loop system

$$M(x)\dot{v} + C(x, v)v + g(x) = Q^T \tau + J_f^T(x)w_f \tag{16a}$$

$$\ddot{\tilde{\tau}} + D_\tau \dot{\tilde{\tau}} + \left(K_\tau + KB^{-1}\right)\tilde{\tau} = 0 \tag{16b}$$

$$B\ddot{\eta} + K_H \tilde{H} K \left(\bar{q}_y(\eta) - q\right) + D_\eta \dot{\eta} + K_\eta \tilde{\eta} = 0 , \tag{16c}$$

where, the input function τ_d will be designed through an optimization based approach similar to the one used in Sect. 3. Notice also that

$$Q^T \tau = Q^T K \left(\eta - q\right) + Q^T \tau_d + Q^T \tilde{\tau} . \tag{17}$$

The details of the optimization problem will be presented later on. For now assume that τ_d has been computed and consider the system conditionally to the convergence of the torque error, i.e. $\tilde{\tau} = \mathbf{0}$. Therefore, the system to consider for the design of τ_d is

$$M(x)\dot{v} + C(x,v)v + g(x) - Q^T K\left(\eta - q\right) = Q^T \tau_d + J_f^T(x)w_f \qquad (18a)$$

$$B\ddot{\eta} + K_H \tilde{H} K\left(\bar{q}_y(\eta) - q\right) + D_\eta\dot{\eta} + K_\eta\tilde{\eta} = \mathbf{0} . \qquad (18b)$$

In a first step, it is assumed that $Q^T \tau_d$ can be freely chosen and therefore, mutatis mutandis, from (6) and (14) it follows

$$Q^T \tau_d = \bar{J}_N^T \left(\bar{\Gamma}\zeta + \begin{bmatrix} Z_1\left(g - Q^T K\left(\eta - q\right) + J_1^T \phi_{d_1}\right) \\ \vdots \\ Z_{r-1}\left(g - Q^T K\left(\eta - q\right) + J_{r-1}^T \phi_{d_{r-1}}\right) \\ 0 \end{bmatrix} \right) . \qquad (19)$$

Compared to (6), one notices that there is no compensation of gravitational and elastic torques for the last task, i.e. the task responsible for the generation of the limit cycle. This is not surprising since the total potential energy is directly considered for the generation of the limit cycle. In (19), each ϕ_{d_i} is a PD term that guarantees the regulation of the task coordinates y_i to their desired value y_{d_i}. Both y_i and y_{d_i} are still to be defined in order to make the robot jump.

The exact definition of the constraints is exemplified, in the next subsection, for the robot in Fig. 1. Nevertheless, the choice is always based on similar tasks as in Sect. 3, i.e. a task for the end effectors in contact with the environment, one for those free to move, one for the CoM and one for the posture. In particular, the choice of having the end effectors in contact with the environment as the first task is key to the design process. Because of this choice, $J_1 = J_f$ and it follows that the reaction wrench will only appear in the first task when performing the change of coordinates used in Sect. 2. This is a consequence of the property

$$J_1(q)\bar{J}_N^{-1}(q) = \begin{bmatrix} E & O & \dots & O \end{bmatrix} ,$$

where E is the identity matrix and O a matrix of zeros. Therefore, as in [10], during the remainder of the design phase, the external wrenches are not taken into account.[4]

The last step of the design process consists in redistributing ϕ_d via an optimization problem as in Sect. 3. In this way, the restrictions on the admissible $Q^T \tau_d$ as well as the contact constraints can be taken into account. In particular, (19) can be rewritten

---

[4]This is justified by the fact that the optimization realizes a wrench at the feet, which does not violate the constraints of keeping them on the floor. If this is the case, the wrench w_f will then be the reaction to the wrench exerted by the feet.

Fig. 6 The weights in the cost function are switched depending on the contact phase. The tasks are defined in Sect. 6.1

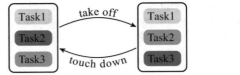

as

$$
\begin{bmatrix} \mathbf{0} \\ \boldsymbol{\tau}_d \end{bmatrix} = \begin{bmatrix} \boldsymbol{\Xi}_u \\ \boldsymbol{\Xi}_a \end{bmatrix} \boldsymbol{\phi}_d + \underbrace{\bar{\boldsymbol{J}}_N^T \left(\bar{\boldsymbol{\Gamma}} \boldsymbol{\zeta} + \begin{bmatrix} \boldsymbol{Z}_1 \boldsymbol{g} - \boldsymbol{Q}^T \boldsymbol{K} \left(\boldsymbol{\eta} - \boldsymbol{q} \right) \\ \vdots \\ \boldsymbol{Z}_{r-1} \boldsymbol{g} - \boldsymbol{Q}^T \boldsymbol{K} \left(\boldsymbol{\eta} - \boldsymbol{q} \right) \\ \boldsymbol{0} \end{bmatrix} \right)}_{\left[\boldsymbol{b}_u^T \ \boldsymbol{b}_a^T \right]^T} . \tag{20}
$$

Finally, $\boldsymbol{\tau}_d$ can be obtained solving the following optimization problem

$$
\min_{\boldsymbol{\phi}} \left(\boldsymbol{\phi} - \boldsymbol{\phi}_d \right)^T \boldsymbol{W} \left(\boldsymbol{\phi} - \boldsymbol{\phi}_d \right) \tag{21a}
$$

$$
\text{s.t. } \boldsymbol{\Xi}_u \boldsymbol{\phi} + \boldsymbol{b}_u = \boldsymbol{0} \tag{21b}
$$

$$
f_{min,i} \leq f_{i,\perp} \tag{21c}
$$

$$
\| \boldsymbol{f}_{i,\|} \| \leq \mu_i f_{i,\perp} \tag{21d}
$$

$$
CoP_i(\boldsymbol{f}_i) \in \mathcal{S}_i \tag{21e}
$$

$$
\boldsymbol{\tau}_{min} - \boldsymbol{b}_a \leq \boldsymbol{\Xi}_a \boldsymbol{\phi} \leq \boldsymbol{\tau}_{max} - \boldsymbol{b}_a \tag{21f}
$$

and setting $\boldsymbol{\tau}_d = \boldsymbol{\Xi}_a \boldsymbol{\phi}^* + \boldsymbol{b}_a$, where $\boldsymbol{\phi}^*$ is the optimal solution.[5]

Concerning the flight phase, it is important to remember that the CoM dynamics cannot be influenced by the motors. Similarly, not all the velocity coordinates can be freely influenced since they have to satisfy the conservation of the angular momentum [4]. The strategy consists, therefore, in switching the weights of the different tasks used in the cost function. A graphical representation of how the weights are updated for the C-Runner can be found in Fig. 6. Notice that while the number associated to the task defines how they are organized in the task hierarchy, i.e. defines the inertial decoupling of the tasks, the weights define which forces the optimization is more likely to modify from the desired values in order to satisfy the constraints of the problem.

---

[5]The time differentiation of the signal $\boldsymbol{\tau}_d$ can be obtained by filtering techniques.

6.1 The C-Runner

The C-Runner, Fig. 1, is a two-legged planar testbed with linear series elastic actuation. The robot is equipped with torque and position sensors, as well as force sensors into each silicone half domes representing the contact points at the feet, to provide a full state measurement. The trunk is connected to a boom, which constrains the robot to move in the sagittal plane. Therefore, the link-side configuration of the robot is defined by nine parameters, three for the trunk and one for each joint. More details on the robot can be found in [11]. In order to achieve the desired jumping pattern, the first task is chosen to define the pose of the feet (6 coordinates), the second the horizontal position of the CoM and the trunk orientation (2 coordinates), the third the vertical position of the CoM (1 coordinate). A unique link-side configuration exists, which fulfills all the tasks.

Task one (feet pose) This task is responsible for generating the required interaction wrenches. The feet are asked to stay parallel to the floor and in contact with the ground. The weights of this task are the lowest in order to allow the optimization to use the feet for providing the necessary reaction wrenches by pushing into the floor. In Sect. 7, it is discussed the necessity to have a precise foot placement for gaits like running and walking. In these cases, the weights of this task have to be the highest during the flight phase.

Task two (CoM horizontal position and trunk orientation) The position of the CoM is responsible for the balancing capabilities of the robot. The horizontal position of the CoM is chosen to be in the middle of the support polygon. The weights of this tasks are the highest during the contact phase, as it is fundamental to prevent the robot from falling. During the flight phase, instead, the weights are lowered since only the relative position between the CoM and the feet can be influenced. Finally, the trunk orientation completes the overall posture of the robot.

Task three (CoM vertical position) This task is responsible for the oscillatory behavior of the sequence of jumps. Assuming that all the other tasks are perfectly satisfied, it follows that the floor will exert a net force on the robot which is parallel to the weight. This is justified by the fact that a movement of the CoM along the horizontal position would be otherwise produced. Therefore, the vertical oscillation will continue until the energy is high enough to lead to lift off, similarly to Sect. 5.

6.2 Evaluation

The proposed control law is evaluated in an experiment, for which the values of the proportional action for each PD term are listed in Table 1, while the correspondent coefficient of the derivative action was always chosen using a damping ratio of 0.2.

Starting from the initial configuration shown in Fig. 1, corresponding to the equilibrium, the desired energy is set to 10J and then to zero, as shown in Fig. 7. The plot also shows the real value of the energy. Concerning the evolution of the energy,

Table 1 Coefficients for the proportional action of the PD terms

	Task one	Task two
Linear	$600\ \frac{N}{m}$	$2000\ \frac{N}{m}$
Angular	$80\ \frac{Nm}{rad}$	$800\ \frac{Nm}{rad}$

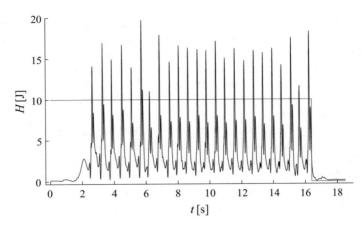

Fig. 7 Desired energy (blue) and actual value (black)

oscillations were expected if for no other reason than the presence of the impacts, as seen in Sect. 5. It should be mentioned that there is no special requirement of reaching the exact desired energy, as long as the desired jumping behavior is obtained. If a higher jumping height is requested, one might increase the desired value of energy and/or the gain K_h, within the limits set by the hardware.

The resulting jumping height of the CoM is reported together with the height of one of the feet in Fig. 8. As it can be noticed, the robot starts from the equilibrium configuration, then the jumps are performed according to the desired energy and finally stops again when the energy goes back to zero.

Finally, in Fig. 9 one can see three entries of the inputs and outputs of the optimization problem. All the desired values of the PD terms are well tracked when the robot is at a rest or making small oscillations, except for the vertical force of the foot. The mismatch is actually the sign that the optimization is working properly, although one might think exactly the opposite. When the robot is at rest on the floor, the desired PD term is zero since the foot is exactly stopped at the required height. Nevertheless, the optimization "knows" that the gravity needs to be compensated and, as the foot task has a low weight in the cost function, commands the robot to push into the floor. On the other hand, when the robot starts jumping, deviations can be observed also for the horizontal position of the CoM. This is again a consequence of the constraints and how the weights are chosen in Fig. 6. The touch down events can be easily recognized by the jumps in the plots.

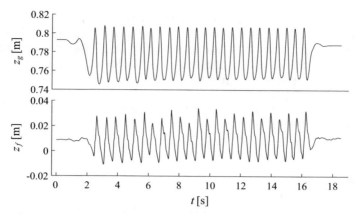

Fig. 8 Vertical position z_g of the CoM and height z_f of one of the feet

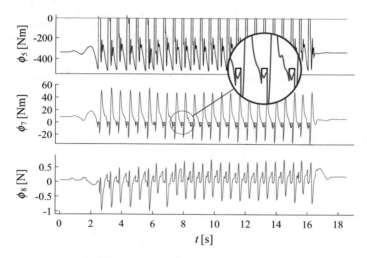

Fig. 9 Values of ϕ for the tasks: left foot height, trunk orientation and CoM horizontal position (from top to bottom). The desired values are in blue and the real in black

7 Conclusion

Our previous controller for limit cycle generation was incorporated in an optimization framework (originally designed for balancing), in order to realize a jumping pattern. A toy model was used to motivate the use of energy regulation to achieve the desired jumps, showing that energy regulation is not beneficial just to exploit the presence of the elastic joints. The controller was evaluated in an experiment using a planar elastically actuated biped robot, which shows the effectiveness of the controller and a behavior similar to the toy example. This controller can be used as core for developing additional locomotion patterns. Crucial in these cases becomes the role of planning and the presence of a state machine. The position of the feet not in contact with the floor is, in fact, paramount for a successful stride and the state machine has to orchestrate the sequence of different phases, i.e. double support, single support and flight phase. These topics are part of future works.

Acknowledgment The authors would like to thank Alexander Werner and Florian Loeffl for their constant support with the hardware.

References

1. Canudas-de-Wit, C., Siciliano, B., Bastin, G.: Theory of Robot Control. Springer, London (1996)
2. de Luca, A.: Dynamic control of robots with joint elasticity. In: IEEE International Conference on Robotics and Automation (ICRA), pp. 152–158. Philadelphia, USA, Apr 1988
3. Garofalo, G., Englsberger, J., Ott, C.: On the regulation of the energy of elastic joint robots: excitation and damping of oscillations. In: American Control Conference (ACC), pp. 4825–4831. Chicago, USA, Jul 2015
4. Garofalo, G., Henze, B., Englsberger, J., Ott, C.: On the inertially decoupled structure of the floating base robot dynamics. In: 8th Vienna International Conference on Mathematical Modelling, pp. 322–327. Vienna, Austria, Feb 2015
5. Garofalo, G., Ott, C.: Limit cycle control using energy function regulation with friction compensation. IEEE Robot. Autom. Lett. (RA-L) **1**(1), 90–97 (2016)
6. Garofalo, G., Ott, C.: Energy based limit cycle control of elastically actuated robots. IEEE Trans. Autom. Control **62**(5), 2490–2497 (2017)
7. Garofalo, G., Ott, C., Albu-Schäffer, A.: Walking control of fully actuated robots based on the bipedal SLIP model. In: IEEE International Conference on Robotics and Automation (ICRA), pp. 1999–2004. Saint Paul, USA, May 2012
8. Garofalo, G., Ott, C., Albu-Schäffer, A.: Orbital stabilization of mechanical systems through semidefinite Lyapunov functions. In: American Control Conference (ACC), pp. 5735–5741. Washington DC, USA, Jun 2013
9. Gehring, C., Coros, S., Hutter, M., Bloesch, M., Hoepflinger, M.A., Siegwart, R.: Control of dynamic gaits for a quadrupedal robot. In: IEEE International Conference on Robotics and Automation (ICRA), pp. 3287–3292. Karlsruhe, Germany, May 2013
10. Henze, B., Dietrich, A., Ott, C.: An approach to combine balancing with hierarchical wholebody control for legged humanoid robots. IEEE Robot. Autom. Lett. (RA-L) **1**(2), 700–707 (2016)
11. Loeffl, F.C., Werner, A., Lakatos, D., Reinecke, J., Wolf, S., Burger, R., Gumpert, T., Schmidt, F., Ott, C., Grebenstein, M., Albu-Schäffer, A.: The DLR C-Runner: concept, design and experiments. In: IEEE/RAS International Conference on Humanoid Robots, pp. 758–765. Cancun, Mexico, Nov 2016
12. Ott, C., Dietrich, A., Albu-Schäffer, A.: Prioritized multi-task compliance control of redundant manipulators. Automatica **53**, 416–423 (2015)
13. Ott, C., Kugi, A., Nakamura, Y.: Resolving the problem of non-integrability of nullspace velocities for compliance control of redundant manipulators by using semi-definite Lyapunov functions. In: IEEE International Conference on Robotics and Automation (ICRA), pp. 1456–1463. Pasadena, USA, May 2008
14. Park, H.W., Park, S., Kim, S.: Variable-speed quadrupedal bounding using impulse planning: Untethered high-speed 3D running of MIT Cheetah 2. In: IEEE International Conference on Robotics and Automation (ICRA), pp. 5163–5170. Seattle, USA, May 2015

15. Pratt, G.A., Williamson, M.M.: Series elastic actuators. In: IEEE/RSJ International Conference on Intelligent Robots and Systems (IROS), pp. 399–406. Pittsburgh, USA, Aug 1995
16. Sakka, S., Yokoi, K.: Humanoid vertical jumping based on force feedback and inertial forces optimization. In: IEEE International Conference on Robotics and Automation (ICRA), pp. 3752–3757. Barcelona, Spain, Apr 2005
17. Siciliano, B., Sciavicco, L., Villani, L., Oriolo, G.: Robotics: Modelling Planning and Control. Springer Publishing Company, Incorporated (2008)
18. Spong, M.W.: Modeling and control of elastic joint robots. ASME J. Dyn. Syst. Meas. Control **109**, 310–318 (1987)
19. Wolf, S., Eiberger, O., Hirzinger, G.: The DLR FSJ: Energy based design of a variable stiffness joint. In: IEEE International Conference on Robotics and Automation (ICRA), pp. 5082–5089. Shanghai, China, May 2011

Multifunctional Principal Component Analysis for Human-Like Grasping

Marco Monforte, Fanny Ficuciello and Bruno Siciliano

Abstract In this paper, a method to derive the synergies subspace of an anthropomorphic robotic arm–hand system is proposed. Several human demonstrations of different objects grasping are measured using the Xsens MVN suite and then mapped to a seven Degree-of-Freedom (DoF) robotic arm. Exploiting the anthropomorphism of the kinematic structure of the manipulator, two Closed-Loop Inverse Kinematics (CLIK) algorithms are used to reproduce accurately the master's movements. Once the database of movements is created, the synergies subspace are derived applying the Multivariate Functional Principal Component Analysis (MFPCA) in the joint space. A mean function, a set of basis functions for each joint and a pre-defined number of scalar coefficients are obtained for each demonstration. In the computed subspace each demonstration can be parametrized by means of a few number of coefficients, preserving the major variance of the entire movement. Moreover, a Multilevel Neural Networks (MNNs) is trained in order to approximate the relationship between the object characteristics and the synergies coefficients, allowing generalization for unknown objects. The tests are conducted on a setup composed by a KUKA LWR 4+ Arm and a SCHUNK 5-Finger Hand, using the Xsens MVN suite to acquire the demonstrations.

1 Introduction

In order to execute complex tasks and interact deftly with the environment, robots are becoming more and more sophisticated and endowed with great manipulation capabilities. This dexterity requires a high number of DoFs of the arm–hand system, along with high sensorimotor and reasoning skills. On the other hand, this implies the need of greater computational capabilities, more powerful hardware and different control strategies to speed up and simplify the execution of these complex tasks.

M. Monforte (✉) · F. Ficuciello · B. Siciliano
PRISMA Lab, Department of Electrical Engineering and Information Technology,
Università degli Studi di Napoli Federico II, Via Claudio 21, 80125 Naples, Italy
e-mail: marco.monforte@iit.it

© Springer International Publishing AG, part of Springer Nature 2019
F. Ficuciello et al. (eds.), *Human Friendly Robotics*, Springer Proceedings
in Advanced Robotics 7, https://doi.org/10.1007/978-3-319-89327-3_4

47

Despite a classical model-based method [1], in the latest years robotics and neuroscience researchers have focused their efforts on bio-inspired systems and joined their efforts in order to try to understand and recreate particular aspects, structures and/or behaviours of biologic systems. One of the main sources of inspiration is, of course, the human being [2, 3].

Santello et al. [4] and later Mason et al. [5] show that human hand movements during grasping tasks are computed in a space of highly reduced dimensions with respect to the joint space.

This reduced subspace is called synergies subspace. A synergy mapping method from a human hand to a robotic one is proposed in [6], operating in the task space and with the usage of a virtual sphere. In [7] a Neural Network (NN) is trained with the object's features and the synergies coefficients in order to generalize the grasping strategy.

Grasping tasks do not rely only on the hand, but also on the arm configuration. A possible way to generalize the arm motion could be the Dynamical Movement Primitives (DMPs) [8], as explored in [9, 10]. Another possibility is the MFPCA. With the aim of extending the work done in [7] and in [3], we propose to obtain the synergies subspace of a robotic arm–hand system with the MFPCA technique, which is an extension of the Functional Principal Component Analysis (FPCA) to the multivariate case. FPCA, that allows computing the dominant modes of variation in functional data, can be seen as the extension of the Principal Component Analysis (PCA) to the functional case. The motivation to investigate the MFPCA for dimensionality reduction is to preserve as much information as possible contained in the motion coordination. Instead of final configuration considered in [3, 7], here the demonstrations are time series.

To this purpose, reach-to-grasp movements are acquired by teleoperating the robotic system. The grasped objects have different dimensions. On this set of demonstrations, the MFPCA is applied in order to link each object to a fixed number of scalar coefficients that reproduce the grasping movement.

Finally, MNNS are trained with pairs of object's characteristics and scalar coefficients in order to try to generalize the behaviour to new objects. A schematic recap is in Fig. 1. The scope of this work is to compute the synergies subspace of the arm–hand system and to evaluate if MNNs, successfully tested on the hand, are good enough to find the function that maps the object characteristics into the MFPCA coefficients. In future works, a Reinforcement Learning (RL) approach could be introduced to explore the space around the final configuration computed by the MNN.

The paper is organized as follows: Sect. 2 describes the experimental setup; Sect. 3 describes the technical approach adopted for the human motion mapping on the robotic arm, for the synergies subspace derivation and for the MNN training; Sect. 4 illustrates the mapping procedure using the Xsens MVN; in Sect. 5 the MFPCA is briefly described together with the synergies subspace computation procedure; in Sect. 6 the training of the MNN is analyzed along with the results; finally, Sect. 7 provides conclusions and future developments.

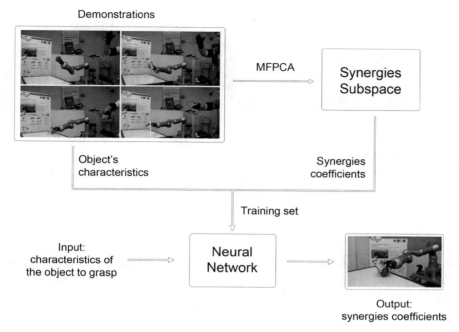

Fig. 1 Diagram of the methodological approach

2 Experimental Setup

The arm–hand robotic system used to acquire the demonstrations is composed by a SCHUNK 5-Finger Hand (S5FH) and a KUKA Lightweight Robot (LWR) 4+ arm (Fig. 2). The S5FH is an underactuated anthropomorphic robotic hand with 20 DoFs actuated by 9 motors. Its mechanical design presents an example of hand synergies through mechanical couplings, to reduce the weight and the dimensions. The KUKA robotic arm has 7 DoFs like the human arm. The redundant joint allows changing the internal configuration of the arm while keeping the end-effector in a fixed pose. The Robot Operating System (ROS) is used to control the S5FH through an SVH Driver developed by Forschungszentrum Informatik (FZI). This driver for the low-level interface allows an easy control using a C++ customized library. The KUKA LWR 4+ is controlled using the Fast Research Interface (FRI) library, developed at Stanford University. The motion capture tool used to acquire the demonstrations is the Xsens MVN package, composed by the Xsens suite and the MVN software. The Xsens suite does not need cameras, emitters or markers but consists essentially of 17 MTx sensors. After a calibration process, it is possible to obtain position and orientation of the body parts where the sensors are attached. Each MTx includes a 3D gyroscope, a 3D accelerometer and a 3D magnetometer. The MVN software, instead, receives the data from the suite and reproduces the human movements in a

Fig. 2 SCHUNK 5-Finger Hand, KUKA LWR 4+ and Xsens suite

simulator. Moreover, the network streamer can be used to send the data received to third applications using UDP/TCP-IP packets. Finally, Matlab NN Toolbox has been used to train the MNNs.

3 Technical Approach

In order to replicate human manipulation skills, artificial limbs for robotics and prosthetics require a biomechanical human-like structure with high number of DoFs. On the other hand, recent advances in neuroscience research have shown that the human Central Nervous System (CNS) coordinates these DoFs to generate complex tasks by means of a synergistic organization at postural, muscular and neural level. Using a limited set of motion patterns, named postural synergies, a wide range of grasping tasks can be achieved, reducing substantially the dimension of the robotic hand control problem. The main objective of this work is to apply this concept of dimensionality reduction to a robotic arm, generalizing it to the entire movement and not only looking at the final pose. We have investigated the MFPCA technique and to this purpose, we firstly need a set of demonstrations. Thus, the first problem is how to create this dataset.

To provide the robotic system with human-like movements, the demonstrations are obtained by teleoperating the systems. The main advantages of this technique is to be fast and general: if we need further demonstrations, we just execute them, recording the variables of interest while performing the task. Imitation learning can be achieved in different ways, the main methodologies are two: teleoperation, using tracking systems like vision, exoskeletons or other wearable motion sensors, and kinesthetic teaching, where the robot is physically moved by the human to execute the task. Since we deal with high-DoF platforms, the latter method is not really practicable: the human master should move too many joints at the same time, resulting in discontinuous movements and jerky trajectories. Therefore, in this work the system is teleoperated using a motion tracking suite. However, in order to reproduce the demonstrations as close as possible, teleoperation needs a mapping of human motion to the robotic system. For this purpose, two Closed-Loop Inverse Kinematics (CLIK) algorithms are used to control the KUKA arm. The first CLIK algorithm controls the first three joints (shoulder) by giving to the robotic arm the same orientation of the master. The second CLIK algorithm controls the S5FH palm orientation with respect to the robotic forearm using the last three joints of the robot. Moreover, the null space of the second CLIK Jacobian, that relates the velocities of the last four joints to the Cartesian space velocities, is used to reproduce the angle between the human arm and forearm. Once the mapping problem is solved, several demonstrations are acquired to cover a complete grasp taxonomy [6].

Robot joint trajectories are recorded during teleoperated reach and grasp tasks of different objects (Table 1). The considered objects are "balls" and "cylinders"of

Table 1 Objects used for the demonstrations

	Object	Diameter [cm]	Height [cm]
Balls	Bottle cap	2.7	2.7
	Table tennis ball	4.4	4.4
	Plastic plum	5.2	5.2
	Racquetball	5.5	5.5
	Plastic peach	5.9	5.9
	Tennis ball	6.4	6.4
	Blue ball	6.9	6.9
	Plastic orange	7.1	7.1
	Plastic apple	7.4	7.4
	Red ball	8.8	8.8
	Softball	9.6	9.6
Cylinders	N° 1	1.2	21
	N° 2	1.8	21
	N° 3	2.4	21
	N° 4	3.3	21
	N° 5	4.9	18.2
	N° 6	5.1	16
	N° 7	5.4	23
	N° 8	6	21
	N° 9	6.6	21
	N° 10	7	21.0
	Chips tube	7.5	21.0

different dimension grasped with precision and/or a power grasp, depending on whether the object allows both or not.

Once the dataset is created, Dynamic Time Warping technique [11] is applied for better data conditioning, and a subsampling is carried out in order to have all the time series as similar as possible and with same length.

Finally, we have a database of 38 demonstrations with the same length. Afterwards, MFPCA technique is applied and a set of scalar coefficients parametrizing the demonstrations are obtained.

The pairs constituted by object characteristics and related synergies coefficients have been used to train an MNN using Matlab Toolbox. The goal is to generalize to novel objects the synergy-based grasping motion obtained from the MFPCA.

4   Human Motion Mapping

Regarding the human hand motion mapping to the S5FH, the work already done in [8] is exploited. A Jacobian matrix of the synergies is computed and used into a CLIK algorithm to map human hand grasping configurations to the robotic hand. Further details of the mapping and the results can be found [9].

To map the human motion to the robotic arm, we recall that the MTx sensor on the Xsens suite provides position and orientation of the human limbs whith respect to the global frame of the mo-cap system. Moreover, the KUKA arm presents 7 DoFs as the human arm. Thus the developed method to map the motion involves two CLIK algorithms that take the arm and the hand orientations as reference input respectively.

We look at the robotic arm like two separate kinematic structures: the first one made by three joints (shoulder) and the second one made by four joints (wrist) with one redundant DoF (elbow). To reproduce at best the human movement, the angle between the arm and the forearm is computed with geometric properties and assigned to the redundant joint, exploiting the null space of the Jacobian matrix of the second kinematic structure.

With the described mapping algorithm, we are able to teleoperate the KUKA arm, as can be seen in Fig. 3. 38 demonstrations are performed and the seven joint trajectories are recorded along the entire task execution.

5   Synergies Subspace Computation

The PCA technique is a tool for multivariate data analysis and dimensionality reduction, extended to functional data from 1950, as Functional Principal Component Analysis (FPCA). Functional data can be seen as the realization of a stochastic process that, under mild assumptions, can be expressed as a sequence of uncorrelated random variables, called Functional Principal Components (FPCs). The i-th realization of the stochastic process X has the form:

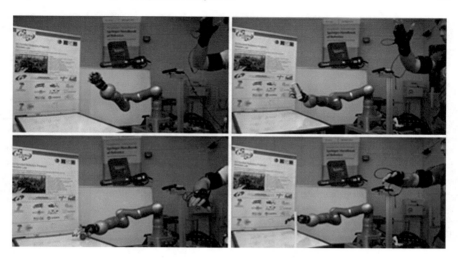

Fig. 3 Snapshots of performed demonstrations

$$X_i(t) = \mu(t) + \sum_{k=1}^{\infty} \xi_{ik}\varphi_k(t) \quad t \in T \tag{1}$$

where $\mu(t)$ is the mean function of the dataset, φ_k is the k-th functional principal component and ξ_{ik} is the k-th coefficient of the respective FPC that parametrize the i-th realization.

The basic idea is to map the initial domain of the dataset into a subspace that allows parametrizing each realization with a reduced number of scalar parameters. The major advantage of these techniques is that they preserve major variance with respect to other transformations like Fourier or Wavelet.

In the PCA each input realization is a vector of scalar components and the dimension is finite. The mean value is a vector and the eigenvalues/eigenvectors of the entire dataset have a finite number of components. In the FPCA each realization is a time series, hence the dimension is infinite. The mean value is a mean function and the number of eigenvalues/eigenfunctions of the entire dataset is infinite (Table 2).

The FPCA can be extended to the multivariate case as Multifunctional Principal Component Analysis (MFPCA). Each observation has now $n \geq 2$ functions, possibly defined on different domains $T_1, \ldots T_n$. It has the form of a vector X such that:

$$X_i(\mathbf{t}) = (X_i^{(1)}(t_1), \ldots, X_i^{(n)}(t_n)) \in \mathbb{R}^n \tag{2}$$

with $\mathbf{t} = (t_1, \ldots, t_n) \in T = T_1 \times \cdots \times T_n$.

The decomposition of a realization of the stochastic process can be thus extended as:

Table 2 Comparison between PCA and FPCA techniques

Element	In PCA	In FPCA
Data	$X \in \mathbb{R}^p$	$X \in L^2(T)$
Dimension	$p < \infty$	∞
Mean	$\mu = \mathrm{E}(X)$	$\mu(t) = \mathrm{E}(X(t))$
Covariance	$\mathrm{Cov}(X) = \Sigma_{p \times p}$	$\mathrm{Cov}(X(s), X(t)) = G(s, t)$
Eigenvalues	$\lambda_1, \lambda_2, \ldots, \lambda_p$	$\lambda_1, \lambda_2, \ldots$
Eigenvectors/eigenfunctions	$v_1, v_2, \ldots, v_p$	$\varphi_1(t), \varphi_2(t), \ldots$
Inner product	$\langle \mathbf{X}, \mathbf{Y} \rangle = \Sigma_{k=1}^{p} X_k Y_k$	$\langle \mathbf{X}, \mathbf{Y} \rangle = \int_T X(t)Y(t)dt$
Principal components	$z_k = \langle X - \mu, v_k \rangle,$ $k = 1, 2, \ldots, p$	$\xi_k = \langle X - \mu, \varphi_k \rangle,$ $k = 1, 2, \ldots$

$$X_i(\mathbf{t}) = \mu(\mathbf{t}) + \sum_{k=1}^{\infty} \xi_{ik}\varphi_k(\mathbf{t}) \quad \mathbf{t} \in T \tag{3}$$

By truncation of the sum to m components, we are capable of parametrizing each realization with m scalar coefficients, also called *scores*.

After the data conditioning, we have now the matrix $\mathbf{D} = \{\mathbf{d}_i \mid i = 1, \ldots, 38\}$ of the demonstrations. Each $\mathbf{d_i}$ is a reaching and grasping execution in different conditions (different objects or same object with different grasp types) and is composed by seven functions representing the values of the KUKA arm joints during the demonstration of the task:

$$\mathbf{d}_i = \left(d_i^{(1)}(t), \ldots, d_i^{(7)}(t) \right) \quad t \in T \tag{4}$$

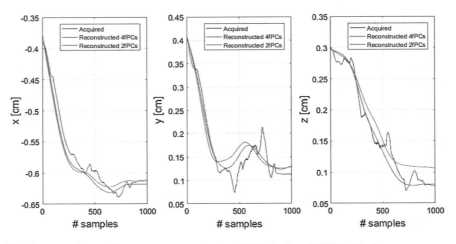

Fig. 4 x, y and z coordinates reconstructed using 2 FPCs (red) and 4 FPCs (blue)

The MFPCA is applied on $\mathbf{D}$ by setting the desired number of FPCs $m = 4$ (see [11] for further details on the procedure). We obtain mean functions $\mu^{(j)}(t)$ and m eigenfunctions $\varphi_1^{(j)}, \ldots, \varphi_m^{(j)}$ for each joint ($j = 1, \ldots, 7$). The m eigenfunctions represent the basis of the subspace for the relative joint. Finally, we obtain a matrix of scores $\Xi \in \mathbb{R}^{38 \times m}$ where each row parametrizes a demonstration. It is important to highlight that the k-th score modulates the correspondent k-th FPC for each joint. In other words, the coefficient the same for all the seven joints, while the eigenfunctions (and thus the basis of the subspace) are different for each of them.

Reconstructing the x, y and z trajectories (Fig. 4) we can see that 2 FPCs represent quite well the original data. Yet, if we rely only on MNNs to grasp the objects, we will need to be more precise. Therefore, the usage of 3 or 4 FPCs will improve performances. Another important advantage is the filtering effect of the MFPCA: the reconstructions eliminate the noise acquired while performing the demonstration.

6 Neural Network Training

To confer autonomy to the system, MNNs can be a suitable choice due to their ability to learn mapping functions. Moreover, the usage of synergies reduces the search space of the learning algorithm, implying also a simplification of the architecture, especially regarding the number of hidden layers and number of neurons.

Two MNNs are used in this work, one for the hand and one for the arm. Their architecture is experimentally chosen by trying different combinations of hidden levels and neurons and analyzing the corresponding performance in terms of Mean Squared Error (MSE). We found out that a feedforward structure with 2 hidden layers and 10 neurons for each is the suitable choice for the particular application.

To learn the function that relates the object shape and grasp type with the arm-hand grasping action, MNNs need a training dataset constituted by pairs of object geometric features and scores obtained from the MFPCA. The input parameters are the *height*, *diameter* and an information on the *grasp type* (power or precision). Since MMNs inputs must have the same dimensions for all the objects, the *height* is set equal to the *diameter* for spherical shaped objects. The output are the three synergy coefficients that determine the configuration of the SCHUNK 5-Finger Hand and the 4 coefficients of the KUKA arm that determine the reaching movement towards the object, respectively for each MNN. It is well known that neural networks performance increase when decrease the dimension of the space where is defined the function to be learned. Therefore, for the arm we decided to test the net with 4 FPCs since they can reconstruct the motion with a lower error with respect to the use of the first two FPCs, meanwhile the dimension of the search space is reduced from 7 to 4 with a good compromise between performance and generalization capabilities. The goal now is to experimentally evaluate if the use of MNNs based on FPCs can generalize grasps of new objects with respect to the database used for training.

The first evaluation is made by comparing the waveforms obtained from the Multilevel Neural Network with the reconstruction shown in Fig. 4 using 4 eigenfunctions.

The comparison is performed on a grasp example contained in the database and is shown in Fig. 5. As we expected, the combination of eigenfunctions computed with the Multilevel Neural Network provides a higher reconstruction error for each grasp in the database. This is the drawback of the generalization process provided by the net. To evaluate the ability of the MNN to generalize the grasps, we have selected two objects with geometrical features closed to objects contained in the database and two objects different from the examples contained in the database.

From experiments represented in Fig. 6, we can state that the MNNs are not good enough to realize grasps of new objects but need to be integrated with strategies based on trial-and-error.

Indeed, from an error analysis performed on a wide set of objects, the average of the errors between the hand palm posture planned by the MNNs and the hand posture measured from the human demonstration is:

$$\mathbf{e}_p = \begin{bmatrix} 0.99 & 0.93 & 1.64 \end{bmatrix} \tag{5}$$

$$\mathbf{e}_o = \left(\begin{bmatrix} 0.65 & 0.51 & 0.55 \end{bmatrix}, 0.20944 \right) \tag{6}$$

where $\mathbf{e}_p$ is the position error expressed in [cm] and $\mathbf{e}_o$ is the orientation error with the axis-angle system (angle in radians).

Despite the pose error that doesn't always ensure successful grasps, the hand for each grasp is very close to the target, this makes the coefficients provided by the MNNs a good initialization for a reinforcement learning algorithm.

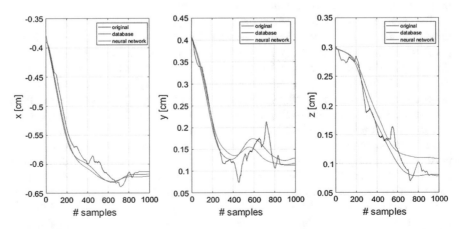

Fig. 5 Comparison between the reconstruction with 4 FPCs of Fig. 4 and the reconstruction with 4 FPCs obtained from the neural network

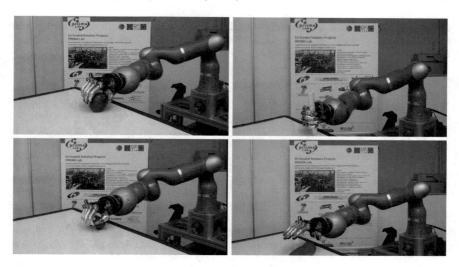

Fig. 6 Examples of grasping actions obtained using the MFPCA outputs. Figures on the top represent grasps of objects closed to ones contained in the dataset. Figures on the bottom represent grasps of objects with dimensions that are more different with respect to the dataset ones

7 Conclusions and Future Work

Multifunctional Principal Component Analysis method to derive the synergies subspace of a robotic system with a high number of DoFs is proposed in this paper. MFPCA technique is an extension of the well-known PCA to the multivariate functional data. Several demonstrations of reach-to-grasp tasks for objects with different dimensions are acquired. The acquisitions are carried out using a motion capture system and a mapping of the human movement to the robotic arm. MFPCA is applied to the collected data by defining a set of FPCs for each joint and a set of scalar coefficients, each one parametrizing a demonstration. By considering the first four FPCs for the arm, it is shown that the reconstructed motion provides a very good approximation of the original functional data.

Afterwords, we used MNNs to generalize the grasps. The training procedure is carried out using the Matlab NN Toolbox and pairs of objects geometric features and relative FPCs coefficients. We have observed that the MNNs it is not sufficient to plan new object grasps. Nevertheless, even if the grasp fails, the hand is close to the correct position. This is a necessary condition for further improvements of the system using a reinforcement learning strategy. Future works, will develop solutions to explore the space around the initial configuration provided by the MNNs, in order to look for a correct final configuration that allows the system to grasp the object.

Acknowledgements The research leading to these results has been partially supported by the RoDyMan project, funded by the European Union (EU) Seventh Framework Programme (FP7/2007-2013) under ERC AdG-320992, and partially by MUSHA project, National Italian Grant under

Programma STAR Linea 1. The authors are solely responsible for the content of this paper, which does not represent the opinion of the EU, and the EU is not responsible for any use that might be made of the information contained therein.

References

1. Ficuciello, F., Carloni, R., Visser, L.C., Stramigioli, S.: Port-Hamiltonian modeling for soft-finger manipulation. In: Proceedings of the IEEE/RSJ Internationl Conference on Intelligent Robots and Systems, Taipei, Taiwan, pp. 4281–4286 (2010)
2. Ficuciello, F., Palli, G., Melchiorri, C., Siciliano, B.: Planning and control during reach to grasp using the three predominant UB Hand IV postural synergies. InL Proceedings of the IEEE International Conference on Robotics and Automation, St. Paul MN, USA, pp. 2255–2260 (2012)
3. Ficuciello, F., Zaccara, D., Siciliano, B.: Synergy-based policy improvement with path integrals for anthropomorphic hands. In: Proceedings of the IEEE/RSJ Internationl Conference on Intelligent Robots and Systems, Daejeon, Korea, pp. 1940–1945 (2016)
4. Santello, M., Flanders, M., Soechting, J.: Postural hand synergies for tool use. J Neurosci 18(23), 10105–10115 (1998)
5. Mason, C., Gomez, J., Ebner, T.: Hand synergies during reach-to-grasp. J. Neurophysiol. 86(6), 2896–2910 (2001)
6. Gioioso, G., Salvietti, G., Malvezzi, M., Prattichizzo, D.: Mapping synergies from human to robotic hands with dissimilar kinematics: An object based approach. In: Proceedings of the IEEE International Conference on Robotics and Automation, Workshop on Manipulation Under Uncertainty, Shanghai, China (2011)
7. Ficuciello, F., Palli, G., Melchiorri, C., Siciliano, B.: Postural synergies and neural network for autonomous grasping: A tool for dextrous prosthetic and robotic hands. In: Proceedings of the International Conference on NeuroRehabilitation, Toledo, Spain (2012)
8. Ijspeert, A.J., Nakanishi, J., Hoffmann, H., Pastor, P., Schaal, S.: Dynamical movement primitives: learning attractor models for motor behaviors. Neural Comput. 25(2), 328–373 (2013)
9. Pastor, P., Hoffmann, H., Asfour, T., Schaal, S.: Learning and generalization of motor skills by learning from demonstrations. In: Proceedings of the IEEE International Conference on Robotics and Automation, Kobe, Japan, pp. 763–768 (2009)
10. Gams, A., Ude, A.: Generalization of example movements with dynamic systems. In: 9th IEEE–RAS International Conference on Humanoid Robots, Paris, France, pp. 28–33 (2009)
11. Giorgino, T.: Computing and visualizing dynamic time warping alignments in R: the dtw package. J. Stat. Softw. 31(7), 1–24 (2009)
12. Ficuciello, F., Federico, A., Lippiello, V., Siciliano, B.: Synergies evaluation of the SCHUNK S5FH for grasping control. In: Proceedings of the 15th International Symposium on Advances in Robot Kinematics, Grasse, France (2016)
13. Jolliffe, I.T.: Principal Component Analysis. Springer (2002)
14. Happ, C., Greven, S.: Multivariate functional principal component analysis for data observed on different (dimensional) domains. J. Am. Stat. Assoc. (2017). https://doi.org/10.1080/01621459.2016.1273115

Part II
Natural Human-Robot Interaction

A General Approach to Natural Human-Robot Interaction

Lorenzo Sabattini, Valeria Villani, Cristian Secchi and Cesare Fantuzzi

Abstract This paper proposes a scheme for letting a human interact with a generic robot in a natural manner. Based on the concept of natural user interfaces, the proposed methods exploit recognition of the users' forearm motion to produce commands for the robotic system. High-level commands are provided based on gesture recognition, and velocity commands are computed for the robot by mapping, in a natural manner, the motion of the user's forearm. The method is proposed in a general manner, and is then instantiated considering two different robotic systems, namely a quadrotor and a wheeled mobile robot. Usability of the system is evaluated with experiments involving users.

Keywords Natural interaction · Human-robot interaction · Human-centered robotics

1 Introduction

In this paper we propose a general methodology for achieving human-robot interaction in a natural manner.

Thanks to recent technological advances in safety and control systems, robots are becoming increasingly common in daily life. Several application domains have

L. Sabattini · V. Villani (✉) · C. Secchi · C. Fantuzzi
Department of Sciences and Methods for Engineering (DISMI), University of Modena
and Reggio Emilia, via Amendola 2, 42122 Reggio Emilia, Italy
e-mail: valeria.villani@unimore.it

L. Sabattini
e-mail: lorenzo.sabattini@unimore.it
URL: http://www.arscontrol.unimore.it

C. Secchi
e-mail: cristian.secchi@unimore.it

C. Fantuzzi
e-mail: cesare.fantuzzi@unimore.it

© Springer International Publishing AG, part of Springer Nature 2019   61
F. Ficuciello et al. (eds.), *Human Friendly Robotics*, Springer Proceedings
in Advanced Robotics 7, https://doi.org/10.1007/978-3-319-89327-3_5

been considered, such as social assistance [1], surveillance [6, 13], tour-guide [14] or floor cleaning [3]. These new application domains pose new challenges, with respect to traditional industrial applications, in which robots are typically utilized by expert users. Conversely, in daily life scenarios, users are often novice to robotics: the design of the interaction system is then extremely relevant, in order to make these applications successful.

A concept that explicitly tackles these issues is that of natural user interfaces (NUIs), which has been developed in the last few years [2, 9, 10]. NUIs encourage a direct expression of mental concepts by intuitively mimicking real-world behavior. Thus, the user is offered a natural and reality-based interaction, which exploits her/his pre-existing knowledge and uses actions that correspond to daily practices in the physical world [10].

This is typically achieved, in NUIs, allowing the user to directly manipulate objects and interact with robots rather than instruct them to do so by typing commands. NUIs try then to overcome the access bottleneck of classical interaction devices such as keyboards, mice and joysticks, by resorting to voice, gestures, touch and motion tracking [2, 9].

In this paper we exploit the concept of NUI to propose a novel hands-free infrastructure-less natural interface for human-robot interaction. It is based on recognizing the motion of the user's forearm to define commands to the robot.

Infrastructure-less means that the proposed interaction system is based on a consumer device (namely, a smartwatch), and does not require any additional dedicated external elements (such as sensors). Using this kind of device brings also additional advantages, providing the user with freedom of movement and letting her/him be immersed in the environment where the robot moves. Thus, the user is able to track the robot with non-fragmented visibility [9].

Considering the aforementioned characteristics, the proposed interaction system can be defined as a *tangible* user interface (TUI) [9], in addition to being a NUI. The concept of TUI is used to describe a large number of interaction systems that rely on a coupling between physical objects and digital information, which is physically embodied in concrete form in the environment. TUIs provide a clear mapping between what the robot does, and what the user expects the robot to do. The proposed interaction system relies on embodied interaction, tangible manipulation, physical representation of data and embeddedness in real space, which are the pillars of TUIs [9].

Preliminary versions of the proposed human-robot interaction system were proposed in [16] for interaction with a quadrotor, and in [15] for interaction with a wheeled mobile robot. In this paper we propose a unified framework, that can be applied to general robotic systems.

The paper is organized as follows. The overall architecture of the proposed human-robot interaction scheme is described in Sect. 2. The architecture is then instantiated into different evaluation scenarios in Sect. 3. Results of the experimental validation are reported in Sect. 4. Finally, concluding remarks are given in Sect. 5.

2   Natural Human-Robot Interaction Scheme

In this paper we propose a general architecture for natural human-robot interaction, whose scheme is depicted in Fig. 1. The proposed scheme is defined as the interconnection of three main entities:

1. the user, who wears a smartwatch (or an activity tracker wristband) which is used to acquire the motion of her/his forearm, and provide her/him with feedback,
2. a data processing system that is in charge of data elaboration,
3. the robot, which receives control inputs from the data processing system.

More in details, the smartwatch acquires characteristic measurements related to the motion of the user's forearm, such as angular displacement, angular velocity and linear acceleration.

These measurements are then elaborated by the data processing system, to implement motion recognition. In particular the motion of the user's forearm is analyzed for:

- recognizing *gestures*, which are used for imposing high-level commands to the robot,
- defining *velocity commands*, by means of a natural mapping between the user's and the robot's motion.

These two different interaction modes will be described in the following subsections.

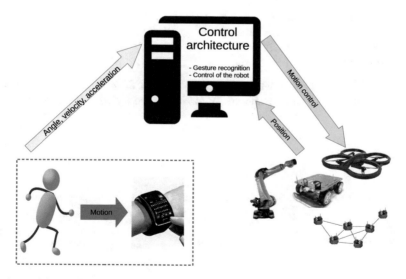

Fig. 1 Overview of the proposed interaction and control system

2.1   Gestures for High-Level Commands

A high-level architecture is implemented, and it allows the user to define different *operational modes* for the robot. An example of such architecture is depicted in Fig. 2, where three main modes are considered:

1. *Autonomous*: in this state, the robot is controlled by means of some pre-programmed algorithm or strategy, and is completely independent from the user. In the simplest case, in this state the robot is stopped.
2. *Supervised*: in this state, the robot implements some semi-autonomous control algorithm, and the user can switch among different algorithms, or provide high-level objectives (e.g. take-off/landing, follow a trajectory, etc.).
3. *Teleoperated*: in this state, velocity commands for the robot are directly computed as a function of the motion of the user's forearm.

The user can change the state of the system by means of *gestures*. A gesture recognition algorithm, based on template matching [7], was proposed in [16].

In this method, data acquired by the accelerometer, gyroscope and magnetometer of the smartwatch are utilized. A template is built, in advance, for each considered gesture: such templates are then utilized, online, comparing them with the motion pattern of the user's forearm. A decision whether a gesture has just occurred is then taken, on-line, by computing suited metrics of comparison.

While, in principle, any set of gestures can be utilized, the following set of gestures was considered:

1. *Up*: sharp movement upwards,
2. *Down*: sharp movement downwards,
3. *Circle*: movement in a circular shape,
4. *Left*: sharp movement to the left,
5. *Right*: sharp movement to the right.

The choice of these gestures was motivated by the fact that they are easy to perform but, at the same time, can be effectively distinguished by common movements of

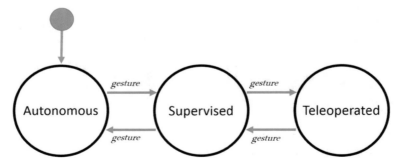

Fig. 2 State machine of the proposed architecture

the forearm. Furthermore, those gestures can be often mapped, in a natural manner, into high-level commands for the robot (e.g. *Up* gesture to control a quadrotor to take-off).

Further details on the gesture recognition algorithm and its validation can be found in [16].

2.2 Natural Mapping Between User's and Robot's Motion

In the teleoperated state (see state machine depicted in Fig. 2) the motion of the forearm of the user is measured by means of the smartwatch tied at her/his wrist and it is mapped into a velocity command for the robot.

Such a mapping is defined in a *natural* manner, and adheres to the principle of NUIs, as defined in Sect. 1, providing the user with a clear mapping between motion of the forearm and expected motion of the robot.

While the exact mapping between the motion of the user's forearm and the velocity command for the robot depends on the specific operational scenario under consideration (e.g. a flying robot and a ground robot move in a completely different manner), some general concepts are depicted in Fig. 3. In particular:

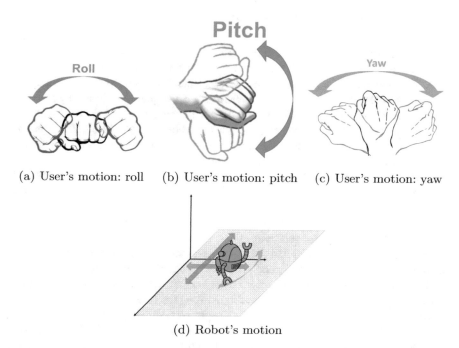

(a) User's motion: roll   (b) User's motion: pitch   (c) User's motion: yaw

(d) Robot's motion

Fig. 3 Mapping between user's and robot's motion

- changing the *roll* angle of the forearm (Fig. 3a) is mapped into a left/right motion command for the robot,
- changing the *pitch* angle of the forearm (Fig. 3b) is mapped into a forward/backward motion command for the robot,
- changing the *yaw* angle of the forearm (Fig. 3c) is used for changing the orientation of the robot.

This represents a general natural mapping, which will be specialized for different robotic systems in the next section.

3   Evaluation Scenarios

In this section we consider the interaction architecture introduced in Sect. 2, and we instantiate it for considering different robotic systems.

We will hereafter not consider issues related to motion control and localization, which have been extensively addressed in the literature and are outside the scope of the paper. For further details, the reader is referred to [5, 8] and references therein.

3.1   Interaction with a Quadrotor

In this section we consider the case in which the user is utilizing the smartwatch to interact with a quadrotor, used, for instance, for inspection operations.

As detailed in [16], considering the general state machine depicted in Fig. 2, we consider the case in which, in the *Autonomous* state, the quadrotor is simply stopped.

The user can then use the *Up* gesture to take the system to the *Supervised* state. In this state, the quadrotor takes off, and goes to a hovering condition. The user can go back to the *Autonomous* state (i.e. quadrotor stopped) with the *Down* gesture: in this case, the quadrotor performs a landing maneuver and stops.

From the hovering condition, the user can exploit the *Left* and *Right* gestures to control the robot to perform, in an automated manner, a left-flip or right-flip maneuver, respectively.

As can be intuitively seen, a natural mapping exists between the gestures and the corresponding action of the quadrotor.

The *Circle* gesture is exploited to take the system to the *Teleoperated* state. Following the general scheme described in Sect. 2.2, the roll angle is used to control the motion of the quadrotor along the forward/backward direction, while the *Pitch* angle is used to control the motion of the quadrotor along the left/right direction. Defining ϑ_r, $\vartheta_p \in [-\pi/2, \pi/2]$ as the roll and pitch angle, respectively, and v_x, $v_y \in \mathbb{R}$ as the velocities of the quadrotor along the forward/backward and left/right direction, with positive sign towards the forward and left direction, respectively, then the velocity commands are computed as follows:

$$v_x(t) = K_r \, \vartheta_r(t), \quad v_y(t) = K_p \, \vartheta_p(t) \tag{1}$$

Constants K_r, $K_p > 0$ are defined to ensure that the maximum angle achievable by the user corresponds to the maximum possible velocity of the quadrotor.

The yaw angle of the smartwatch, namely $\vartheta_y \in [-\pi, \pi]$, is utilized to change the orientation of the quadrotor. This is achieved imposing a setpoint for the yaw angle of the quadrotor, namely $\varphi \in [-\pi, \pi]$. The yaw rate of the quadrotor is then controlled as follows:

$$\dot{\varphi}(t) = K_y \left(\vartheta_y(t) - \varphi(t) \right) \tag{2}$$

where $K_y > 0$ is an arbitrarily defined constant.

The *Circle* gesture can then be utilized for taking the system back to the *Supervised* state.

3.2 Interaction with a Wheeled Mobile Robot

In this section we consider the case in which the user is utilizing the smartwatch to interact with a wheeled mobile robot, for instance for exploring the environment.

In particular, we consider a mobile robot with differential drive kinematics: this choice is motivated by the fact that this kind of robot is quite common in several applications, and its simple kinematic structure allows to keep the notation simple. Nevertheless, the proposed methodology can be easily extended to consider more complex kinematic and dynamic models.

Similar to the quadrotor case, as detailed in [15], considering the general state machine depicted in Fig. 2, we consider the case in which, in the *Autonomous* state, the mobile robot is stopped.

We consider a simplified version of the general state machine depicted in Fig. 2: in particular, in this case, we are not utilizing the *Supervised* state, since we are not considering any pre-programmed control strategy for the robot. However, the proposed control and interaction architecture can be extended, considering high-level behaviors such as trajectory following.

In the proposed architecture, as in the quadrotor case, the user exploits the *Circle* gesture to take the system to the *Teleoperated* state. Define now v, ω to be the linear and angular velocities of the robot, respectively. In this case, the *Roll* angle ϑ_r of the smartwatch is used to control the linear velocity of the robot, as follows:

$$v = K_r \, \vartheta_r \tag{3}$$

where $K_r > 0$ is a constant defined in such a way that the maximum angle that is achievable with the motion of the wrist corresponds to the maximum linear velocity of the mobile robot.

Analogously, the *Pitch* angle ϑ_p is used to control the angular velocity of the robot, namely:

$$\omega = K_p \vartheta_p \tag{4}$$

where the constant $K_p > 0$ is defined such that the maximum angle achievable with the motion of the wrist corresponds to the maximum angular velocity of the robot.

The *Circle* gesture can then be utilized for taking the system back to the *Autonomous* state.

4 Experimental Validation

The proposed smatwatch-based natural interaction system was experimentally validated involving several users. Experiments mainly aimed at evaluating the usability of the proposed interaction system, compared to traditional ones, in different operational scenarios.

For ease of implementation, all the experiments were carried out utilizing an external computer for implementing the high-level control architecture and the gesture recognition. The architecture was implemented in ROS [11]. Specifically, the user wears the smartwatch at her/his wrist, and the motion of her/his forearm is acquired. In particular, velocities, accelerations and angles are measured, and sent from the smartwatch to an external computer, via Wi-Fi communication. The computer is then in charge of processing the received data, computing accordingly the control input, and sending it to the robotic system.

The interaction system was first tested utilizing a Parrot AR.Drone 2.0 quadrotor. In this test, the usability of the proposed interaction system was assessed comparing it with the official smartphone application AR.FreeFlight[1] to control the quadrotor.

The quadrotor was equipped with four needles, and the user was requested to pop some balloons by controlling the motion of the quadrotor, with the two interaction modes: a snapshot of the experiment is shown in Fig. 4. In order to quantify the effectiveness and the usability of the interaction system, we measured the average time to pop a balloon (measured starting from the quadrotor take-off). The experiment involved two subjects, who were asked to pop 10 balloons per method (i.e. with the smartwatch and with the smartphone application). Both the users reported that controlling the motion of the quadrotor with the smartwatch was much more intuitive than with the smartphone application, which caused an increase in the cognitive burden. This is corroborated by quantitative results: the use of the smartwatch resulted in an average time to pop a balloon of 19.9 s, whereas piloting with the application resulted in an average time of 47.6 s. Figure 5 reports the time to pop of both the piloting modalities, for each trial (red solid for the smartwatch, blue dashed for the application), sorted in ascending order. The figure shows the fact that, in 4 trials over 10, the use of the smartphone application leads to a time greater than the greatest time (black dashed line) achieved with the smartwatch.

---

[1] http://www.parrot.com/usa/apps/.

Fig. 4 Assessment of the usability of the proposed architecture

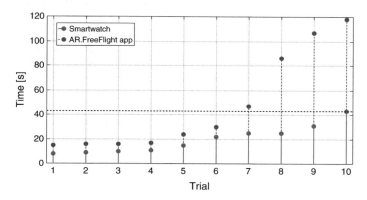

Fig. 5 Time to pop a balloon when piloting the quadrotor with the smartwatch (red solid) and the official application (blue dashed)

The interaction system was then evaluated utilizing a Pioneer P3-AT mobile robot. In this test we aimed at evaluating the benefit of the embeddedness in real space enabled by the use of the smartwatch. For this purpose, the proposed interaction system was compared to a simple remote control system implementing unilateral teleoperation. For this purpose, we used the Geomagic Touch device: the user can move the end-effector of the device, and the motion is translated into a velocity command for the mobile robot. The experiments included 13 users, who were asked to drive the robot through a cluttered environment, using both interaction modes. In particular, the users had to move the robot in an area of 3.3×9.0 (m^2), where seven plastic pins were placed on the ground. Each user was then asked to drive the robot performing a slalom maneuver, avoiding to touch the pins. For the sake of comparison, we measured the total travel time, introducing a 5 s penalty for each touched pin.

The users performed significantly better with the smartwatch than with the remote control device ($p < 10^{-3}$): while pins are typically avoided with both interaction systems, the smartwatch leads to a reduction of the execution time of approximately 33%. Specifically, the total travel time to complete the task was 70.9 s for the smartwatch-based interaction, and 106.6 s when using the remote control device.

5 Conclusions

In this paper we proposed a natural human-robot interaction scheme, which can be applied to general robotic systems. The proposed method is based on recognizing the users' forearm motion to produce commands for the robotic system: high level commands are defined by recognizing user's gestures, while velocity commands are computed for the robot by mapping, in a natural manner, the motion of the user's forearm.

The interaction method was initially proposed in a general manner, and was then detailed considering different robotic systems, namely a quadrotor and a wheeled mobile robot. Usability of the system was evaluated with experiments involving users: the results clearly show the advantage of the proposed interaction system, if compared with traditional ones.

Future work will aim at studying the possibility of introducing a haptic feedback, in terms of modulated vibration of the smartwatch. Preliminary results were proposed in [16] for the control of a quadrotor. Furthermore, we will implement the proposed interaction method on multi-robot systems, considering different centralized or decentralized control strategies, such as in [4]. Specifically, to cite some possible working scenarios, gestures can be used to select different formation shapes, in the common case of robots moving in a formation; in the *Teleoperated* state, the same velocity command can be broadcast to all the robots, or a leader-followers strategy can be considered. Preliminary results have been presented in [17].

Finally, we aim at including measurements of the cognitive workload of the user while interacting with the robot, and adapting the robot's behavior accordingly, in a scenario of affective robotics [12].

References

1. Bemelmans, R., Gelderblom, G.J., Jonker, P., De Witte, L.: Socially assistive robots in elderly care: a systematic review into effects and effectiveness. J. Am. Med. Dir. Assoc. **13**, 114–120 (2012)
2. Blake, J.: Natural User Interfaces in .NET. Manning (2011)
3. Kang, M.C., Kim, K.S., Noh, D.K., Han, J.W., Ko, S.J.: A robust obstacle detection method for robotic vacuum cleaners. IEEE Trans. Consum. Electron. **60**(4), 587–595 (2014)
4. Cacace, J., Finzi, A., Lippiello, V., Furci, M., Mimmo, N., Marconi, L.: A control architecture for multiple drones operated via multimodal interaction in search & rescue mission. In: IEEE International Symposium on Safety, Security, and Rescue Robotics (SSRR), pp. 233–239. IEEE (2016)

5. Campion, G., Chung, W.: Wheeled robots. In: Siciliano, B., Khatib, O. (eds.) Springer Handbook of Robotics, pp. 391–410. Springer (2008)
6. Di Paola, D., Milella, A., Cicirelli, G., Distante, A.: An autonomous mobile robotic system for surveillance of indoor environments. Int. J. Adv. Robot. Syst. 7(1), 19–26 (2010)
7. Duda, R.O., Hart, P.E.: Pattern Classification and Scene Analysis. Wiley, New York (1973)
8. Feron, E., Johnson, E.N.: Aerial robotics. In: Siciliano, B., Khatib, O. (eds.) Springer Handbook of Robotics, pp. 391–410. Springer (2008)
9. Hornecker, E., Buur, J.: Getting a grip on tangible interaction: a framework on physical space and social interaction. In: Proceedings of the SIGCHI Conference Human Factors in Computing Systems (CHI), pp. 437–446. ACM Press (2006)
10. Jacob, R.J., Girouard, A., Hirshfield, L.M., Horn, M.S., Shaer, O., Solovey, E.T., Zigelbaum, J.: Reality-based interaction: a framework for post-WIMP interfaces. In: Proceedings of the SIGCHI Conference Human Factors in Computing Systems (CHI), pp. 201–210. ACM Press (2008)
11. Quigley, M., Conley, K., Gerkey, B., Faust, J., Foote, T., Leibs, J., Wheeler, R., Ng, A.Y.: ROS: an open-source robot operating system. In: Proceedings of the ICRA Workshop Open Source Software, vol. 3, p. 5 (2009)
12. Rani, P., Sarkar, N., Smith, C.A., Kirby, L.D.: Anxiety detecting robotic system—towards implicit human-robot collaboration. Robotica 22(1), 85–95 (2004)
13. Song, G., Yin, K., Zhou, Y., Cheng, X.: A surveillance robot with hopping capabilities for home security. IEEE Trans. Consum. Electron. 55(4), 2034–2039 (2009)
14. Tomatis, N., Philippsen, R., Jensen, B., Arras, K.O., Terrien, G., Piguet, R., Siegwart, R.Y.: Building a fully autonomous tour guide robot. In: Proceedings of the 33rd International Symposium Robotics (ISR) (2002)
15. Villani, V., Sabattini, L., Riggio, G., Levratti, A., Secchi, C., Fantuzzi, C.: Interacting with a mobile robot with a natural infrastructure-less interface. In: Proceedings of the IFAC 20th World Congress International Federation of Automation Control IFAC. IFAC-PapersOnLine (2017)
16. Villani, V., Sabattini, L., Riggio, G., Secchi, C., Minelli, M., Fantuzzi, C.: A natural infrastructure-less human-robot interaction system. IEEE Robot. Autom. Lett. 2(3), 1640–1647 (2017)
17. Villani, V., Sabattini, L., Secchi, C., Fantuzzi, C.: Natural interaction based on affective robotics for multi-robot systems. In: IEEE (ed) 1st International Symposium on Multi-Robot and Multi-Agent Systems (2017)

Construction of a Cooperative Operation Avatar Robot System to Enhance Collective Efficacy

Takafumi Sekido, Ryosuke Sakamoto, Teppei Onishi and Hiroyasu Iwata

Abstract Since the number of people with dementia in Japan is growing, prevention of dementia or improvement of its symptoms are increasingly required. To address this problem, we constructed a system which reduces risk factors for dementia using a cooperative operation robot and an obstacle course activity for the elderly. This robot and the activity enabled players to communicate effectively and increase collective efficacy. We conducted experiments using the cooperative operation system with eight healthy elderly subjects. The subjects expressed more positive feelings, increased the time spent in conversation, and exercised more by swinging their arms during cooperative operation than during single operation. These results show that the cooperative operation robot and activity we developed helped elderly people to improve their collective efficacy. Therefore, this system may have a role in reduction of risk factors for dementia in the elderly.

Keywords Dementia · Elderly · Robot-assisted therapy · Activity · Cooperative operation · Collective efficacy · Communication

1 Introduction

1.1 Background

The number of elderly people in Japan reached 30 million as of 2012, with aging of the population an increasing trend [1]. The prevalence of dementia among the elderly also continues to increase year by year, and in 2025 it will be over 7 million people [2]. These statistics clearly indicate the necessity for treatment or prevention of dementia in Japan. Furthermore, according to Barnes et al. [3], it is effective for prevention of Alzheimer's Disease to improve exercise shortage, smoking, and

T. Sekido (✉) · R. Sakamoto · T. Onishi · H. Iwata
Waseda University, Tokyo, Japan
e-mail: takafumi-sekido@iwata.mech.waseda.ac.jp

© Springer International Publishing AG, part of Springer Nature 2019
F. Ficuciello et al. (eds.), *Human Friendly Robotics*, Springer Proceedings in Advanced Robotics 7, https://doi.org/10.1007/978-3-319-89327-3_6

depression [3]. Therefore, reducing risk factors for dementia has great significance in dementia prevention.

Patients with dementia develop symptoms such as depression, decreased motivation, and poor self-image. These symptoms cause a decrease in self-efficacy which, according to Bandura, is the conviction that one can successfully execute the behavior required to produce the outcomes [4, 5]. In other words, it is the idea that you can accomplish a certain action that you undertake. Bandura identifies four factors that are important to enhance self-efficacy [5]. "Performance accomplishments" are positive experiences of performing a task and succeeding. "Vicarious experience" entails observation of how other people perform the task. "Verbal persuasion" is the self-suggestion that "I can do it" or encouragement from others. "Emotional arousal" involves changes in physiologic responses such as pulse rate and heartbeat. Bandura states that self-efficacy is created by individuals themselves from these four elements [4, 5], although he notes that self-efficacy through performance accomplishments is the strongest and most stable of the factors [4, 5]. Stajkovic et al. have shown that self-efficacy is directly related to human behavior and increases or decreases depending on the circumstances [6]. For example, anxiety may occur when a person is aware of deterioration of their cognitive and motor functions. Such anxiety leads to a decrease in motivation or self-image, which in turn reduces self-efficacy. On the other hand, performance accomplishments, that is, successful achievements, result in positive emotions and in improvement in the sense of self-efficacy. Individuals with high self-efficacy are highly self-affirmative and approach tasks actively. This leads to good results and reinforces the experience of success.

In a group setting, as self-efficacy decreases, collective efficacy within the group also decreases. Collective efficacy, a psychological concept growing out of that of self-efficacy, is defined by Bandura as a belief shared by the group about the ability to systematically implement the necessary actions in achieving the task [7]. The sense of collective efficacy is influenced by the same four features as those promoting self-efficacy [4, 6, 7]. It is therefore reasonable to assume that factors increasing or decreasing self-efficacy would have a similar effect on collective efficacy. Collective efficacy is also defined by other researchers. Zaccaro et al. defined "a sense of collective competence shared among individuals when allocating, coordinating, and integrating their resources in a successful concerted response to specific situational demands" [8]. In addition, group efficacy, team efficacy and other designations are also used [9, 10]. Various designations are made in this way, but since all of them are in conformity with the definitions of Bandura and Zaccaro et al., it can be said that they have almost common meanings. In this research, we use "collective efficacy" which is generally used.

Figure 1 shows mechanisms involved in collective efficacy [11]. Collective cohesiveness, which is an element that is indispensable to collective efficacy, is a social force that maintains the attractions of members to the group and resists divisions among them [11–15]. Patients with dementia suffer a decline in collective efficacy with an attendant decrease in interest and attention, and deterioration in quality of life. Theoretically, improving self-efficacy and collective efficacy could reverse this

Fig. 1 Mechanisms of collective efficacy [11]

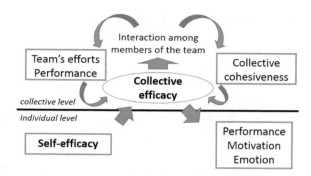

deterioration, leading to amelioration of dementia symptoms, and it might even help prevent elderly dementia.

1.2 Previous Research

There are many kinds of rehabilitation for patients with dementia, such as intelligence training, therapy for functional recovery, drug treatment, and animal therapy [16, 17]. Among them, robot-assisted therapy has attracted attention in recent years.

Animal-type robots such as Paro, a seal robot, and AIBO, a dog robot have been developed. The use of such animal robots yields the same effects as live-animal therapy. Therefore, introduction of these robots is proceeding in hospitals and welfare facilities where animals are prohibited [18–23]. In fact, in hospitals and nursing homes where Paro was introduced, patients are reportedly interested in the robot, and cognitive and motor functions of patients with dementia have improved with contact with Paro [19–22].

Recently, we developed a robot that is manipulated by people performing physical exercise. When patients with dementia did the exercises using this robot, self-efficacy improved in 4 of 4. This shows that robot manipulation by physical exercise increases self-efficacy and is effective against some symptoms associated with dementia [24]. This adds to the evidence from other investigators that social-assistive robots such as humanoid or animal robots may provide effective symptom management.

1.3 Purpose of the Research

The previous studies mentioned have shown that robots are effective as communication tools and the robot manipulation by physical exercise improves self-efficacy in patients with dementia. However, the effects of RAT on collective efficacy among such patients has not been investigated. Our aim is to improve the collective efficacy of elderly people, including patients with dementia. We hypothesized that collective efficacy will increase by having subjects cooperate in an activity using physical exercise to manipulate a robot. In this study, we addressed the following goals:

- Development of an obstacle course race with two people cooperatively operating a robot designed as an avatar.
- Development of an avatar robot for the activity.
- Development of cooperative manipulation system using physical exercise to move the robot.

We conducted tests using the system among healthy elderly people to test the hypothesis that this activity would improve collective efficacy.

2 Methods

2.1 Activity

Factors that enhance the sense of collective efficacy are self-efficacy, efforts within the team, and collective cohesiveness (Fig. 1) [11]. A successful experience enhances self-efficacy [4, 5]. Furthermore, according to previous research, collective cohesiveness is closely related to intimacy, a clearly defined role, and communication [14]. In other words, factors such as successful experiences, efforts within the team, intimacy, role, and communication are important to enhance the sense of collective efficacy. Based on these factors, we set the following three conditions for the activity.

- Players must manipulate the robot using their own motor function (yielding a successful experience).
- Players must make efforts according to the role they are given (efforts within the team, role).
- Players must cooperate and communicate with each other (intimacy, communication).

Analogous to the cooperation needed in a three-legged race, the system we designed required players to work together to move the robot around an obstacle course. The field for the activity contained check points and many obstacles. The two players cooperated by operating one robot together. They moved the robot from the starting point around the check points, avoiding obstacles, and finally returned the robot to the starting position, resulting in a goal. The time from the start to the goal was measured, with players aiming for the fastest time to reach the goal. Each player was given a specific role, one to move the robot forward and the other to rotate it. In the process, players were able improve such factors as their effort and collective cohesiveness.

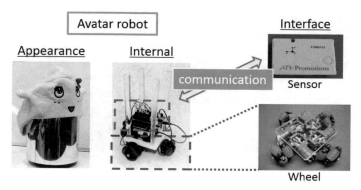

Fig. 2 Avatar robot

2.2 Avatar Robot for the Activity

To carry out the activity, we developed an avatar robot as shown in Fig. 2, with two main specifications, affinity design and omnidirectional mobility. Many robots simulating the appearance of animals, small children, and the like have been developed in the past. It was thought that elderly people would feel greater affinity for a robot wearing a costume.

With regard to mobility, the robot had to move freely in various directions and be able to avoid obstacles. For this reason, we used an omni-wheel at the bottom of the robot. It had four motors, each facing in a different direction. The robot could be freely moved by alternately going straight or turning.

2.3 Simple Physical Exercise Operating System

For many older people and patients with dementia, a simple means of physical operation is preferable. With age, there is often weakening of physical function. A common order in which physical function declines in the elderly is the fingers or hands, lower body, and then upper body parts such as the neck and arms. For this reason, coarse upper body movements are relatively easy to perform. Therefore, we decided that arm swing would be the best operation method. Players wore a wristband (Fig. 3) incorporating an inertial sensor (TSND151, ATR-Promotions Inc.) with Bluetooth wireless communication capability. The wristband was light enough not to feel burdensome when worn. Figure 4 shows a method of single operation. When the player swings his arm, the robot moves forward. In addition, when the player twists his arm to the left and right, the robot turns in the same direction. Through these actions, the sensor measures the change in acceleration, and the information is transmitted to the robot via a computer. This simple operation method is feasible even for frail elderly individuals to use, yielding the hoped for successful experience.

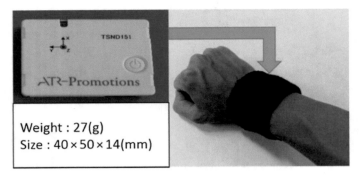

Fig. 3 The operation wrist band and an inertial sensor

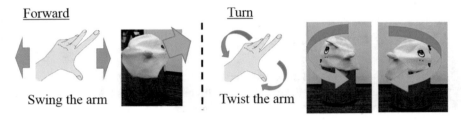

Fig. 4 Single operation method

2.4 Operation by Two People

Figure 5 shows the system. Two wrist bands were prepared, one causing the robot to advance and one causing it to rotate. The players operated the robot by swinging their arms, one advancing the robot and the other rotating it. The players had to figure out how to alternate between advancing and rotating the robot in order to move it to a specific place. This created a situation in which players had to cooperate to operate the robot correctly, which encouraged communication between them.

2.5 Experiment with Elderly Subjects

We conducted experiments in order to investigate the influence of the robot operation system on collective efficacy. The subjects were 8 healthy elderly people aged 60 years or older (mean 66.6 ± 2.87 years old). The intervention was having the subjects operate the avatar robot in an obstacle course race. A group using a cooperative operation system, and a group using a single operation system are set.

Figure 6 shows details of the experiment. The subjects were divided into four pairs. Each group performed the activity cooperatively and with a single operator three times each. For the single operation system, one person operated the robot while

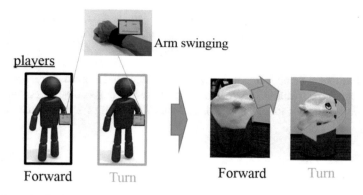

Fig. 5 Players' roles for the cooperative robot operation system

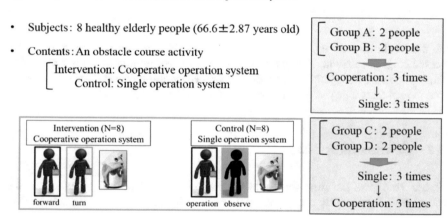

- Subjects: 8 healthy elderly people (66.6±2.87 years old)
- Contents: An obstacle course activity
  - Intervention: Cooperative operation system
  - Control: Single operation system

- Valuables:
  1. Frequency of expression of positive feelings
  2. Time spent conversing
  3. Average acceleration

Fig. 6 Details of the experiment

the other observed alongside. Once the goal was reached, they switched roles, with one operating and one observing. To reduce the influence of the order of operation, two pairs began with cooperative operation and then switched to single operation; the other two pairs reversed that order.

The outcome involved three items: expression of positive feelings, time spent conversing, and average acceleration of the robot (Fig. 6). Items 1, 2, and 3 are highly related to self-efficacy, collective cohesiveness, and efforts within the team, respectively. The overall evaluation was considered to be a measure of collective efficacy. For item 1, AffdexSDK (Affectiva, Inc.) was used. This software has 10,000 images in a database of a person's face expressing emotion and classifies facial expressions in photos and videos into seven types: anger, disgust, fear, joy, sadness, surprise, and contempt. We analyzed the subjects' expressions in video recorded during the experiments, with the frequency of facial expressions of joy counted as positive feelings. For item 2, we measured the amount of time each subject was speaking on the video recording. For item 3, acceleration was calculated from data recorded by the inertial sensors. For statistical analysis, results were compared between cooperative and single operations using an unpaired t-test. The level of significance was set at 5%.

3   Results

3.1   Expression of Positive Feelings

Figure 7a shows the frequency (per minute) of positive feelings during cooperative versus single operation. Positive feelings were expressed significantly more often during cooperative operation than during single operation (p = 0.0002), suggesting that the subjects enjoyed the race more when working cooperatively.

3.2   Conversation Time

Figure 7b shows the time the subjects spent conversing during operation of the robot, with significantly more time spent talking during cooperative operation than

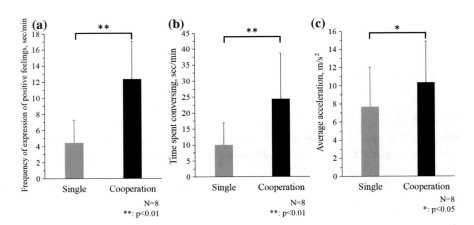

Fig. 7 Results of the experiments

during single operation (p = 0.0021). This result suggests that cooperative operation effectively promoted communication among the players.

3.3 Average Acceleration

As shown in Fig. 7c, the average acceleration during cooperative operation was significantly greater than during single operation (p = 0.0490). This result suggests that subjects were swinging their arm faster during the cooperative operation. That is, it indicates that more effort was made to the arm swinging movement during the cooperative operation.

4 Discussion

In this experiment, we conducted an activity using a robot among eight healthy elderly people and compared their responses when operating the robot cooperatively or individually. We found that that the subjects' positive feelings, time spent in conversation, and average arm acceleration were significantly greater during cooperative operation.

In the case of single operation, there was little communication during the activity, and there were not many smiles. In addition, instead of trying to move their arms faster to make the robot move faster, the operators seemed to be shaking their arms at their own pace. This is because the operator had to perform the activity alone, with the partner not involved in robot operation. On the other hand, since cooperative operation was done by two people, one's success led to the other's success. Therefore, it appeared that they gave instructions back and forth encourage each other to succeed in the activity. In addition, it seems that they tried to swing their arms faster to give their partner the chance to perform the next action quickly. Moreover, when the operators succeeded or failed, the two were talking happily with a smile. In this activity, the consequence of the operation appears in the behavior of the avatar robot. Since the operators feel affinity for the avatar robot, the frustration with respect to the partner was alleviated, leading to the expression of positive emotions.

There were several limitations to this study. Firstly, subjects paired with each other in this experiment were already acquainted. It is necessary to verify the results obtained when a cooperative operation is performed by two people with no previous relationship. The second is the physical ability of the subjects. In this experiment, none of the subjects had any disability with regard to normal physical activity. Whether elderly individuals with limited physical function could successfully use this system will require further investigation with subjects with various levels of functional capacity.

5 Conclusion

We designed an avatar robot with an operating system requiring physical activity by the operators, including a device that allowed cooperative operation to manipulate the robot around an obstacle course and tested it in healthy elderly subjects. As a result of the activity, the subjects had more positive emotions, conversed more, and had better arm swing acceleration when operating the robot cooperatively. These finding indicate that the cooperative operation improved collective efficacy among the elderly subject. Therefore, an activity using a cooperative robot operation system may be useful in reducing risk factors for dementia among the elderly.

Acknowledgements We would like to thank Kazuhiro Yasuda, who is Junior Researcher (Assistant Professor) at our laboratory. In addition, we thank Zenyu Ogawa, who is an external researcher at our laboratory, and his friends for their support on this project.

References

1. Ministry of Internal Affairs and Communications statistics Bureau. http://www.stat.go.jp/data/jinsui/2012np/
2. Cabinet Office, Goverment of Japan. http://www8.cao.go.jp/kourei/whitepaper/w-2016/html/gaiyou/s1_2_3.html
3. Barnes, D.E., Yaffe, K.: The projected impact of risk factor reduction on Alzheimer's disease prevalence. Lancet Neurol. **10**, 819–828 (2011)
4. Bandura, A.: Social foundations of thought and action: a social cognitive theory. Prentice-Hall, Englewood Cliffs (1986)
5. Bandura, A.: Self-efficacy: toward a unifying theory of behavioral change. Psychol. Rev. **84**, 191–215 (1977)
6. Stajkovic, A.D., Luthans, F.: Self-efficacy and work related performance: a meta-analysis. Psychol. Bull. **124**(2), 240–261 (1998)
7. Bandura, A.: Self-efficacy: the exercise of control. Freeman, New York (1997)
8. Zaccaro, S.J., Blair, V., Peterson, C., Zazanis, M.: Collective efficacy. In: Maddux, J.E. (ed.) Self-Efficacy, Adaptation, and Adjustment. The Plenum Series in Social/Clinical Psychology. Springer, Boston, MA (1995)
9. Gibson, C.B.: Do they do what they believe they can? Group efficacy and group effectiveness across tasks and cultures. Acad. Manag. J. **42**(2), 138–152 (1999)
10. Vargas-Tonsing, T.M., Warners, A.L., Feltz, D.L.: The predictability of coaching efficacy on team efficacy and player efficacy in volleyball. J. Sport Behav. **26**(4), 396–407 (2003)
11. Feltz, D.L., Short, S.E., Sullivan, P.J.: Self-efficacy in sport: research and strategies for working with athletes, teams, and coaches. Human Kinetics, Champaign (2007)
12. Hagger, M., Chatzisarantis, N.: The social psychology of exercise and sport. Open University Press, UK (2005)
13. Caron, A.V.: Cohesiveness in sport group: Interpretations and considerations. J. Sport Psychol. **4**, 123–138 (1982)
14. Carron, A.V., Mark, A.E.: Group dynamics in sport. Fitness Information Technology Publishing, West Virginia (2012)
15. Mochida, K., Takami, K., Shimamoto, K.: The effect of individual factors on group cohesiveness and collective efficacy in sports: Effects of members' life skills on the group. Jpn. J. Sports Ind. **25**(1), 25–37 (2015)

16. American Psychiatric Association: Practice guideline for the treatment of patients with Alzheimer's disease and other dementias of late life. Am. J. Psychiatry **154**, 1–39 (1997)
17. Szeto, J.Y., Lewis, S.J.: Current treatment options for Alzheimer's disease and Parkinson's disease dementia. Curr. Neuropharmacol. **14**(4), 326–338 (2016)
18. Rouaix, N., Chavastel, L.R., Rigaud, A.S., Monnet, C., Lenoir, H., Pino, M.: Affective and engagement issues in the conception and assessment of a robot-assisted psychomotor therapy for persons with dementia. Front. Psychology. **8**(1), 1–15 (2017)
19. Shibata, T., Wada, K.: A new approach for mental healthcare of the elderly—a mini-review. Gerontology **57**, 378–386 (2011)
20. Shibata, T.: Therapeutic seal robot as biofeedback medical device: qualitative and quantitative evaluations of robot therapy in dementia care. Proc. IEEE **100**(8), 2527–2538 (2012)
21. Wada, K., Shibata, T., Musha, T., Kimura, S.: Effects of robot therapy for dementia patients evaluated by EEG. In: Proceedings of the IEEE/RSJ International Conference on IROS, pp. 2205–2210 (2005)
22. Hamada, T., Kagawa, Y., Onari, H., Naganuma, M., Hashimoto, T., Yoneoka, T.: Study on transition of elderly people's reactions in robot therapy. In: 11th ACM/IEEE International Conference on Human-Robot Interaction, pp. 431–432. Christchurch (2016)
23. Hamada, T., Okubo, H., Inoue, K., Maruyama, J., Onari, H., Kagawa, Y., Hashimoto, T.: Robot therapy as for recreation for elderly people with dementia—game recreation using a pet-type robot. In: Proceedings of the 17th IEEE International Symposium on Robot and Human Interactive Communication, pp. 174–179. Technische Universität München, Munich, Germany (2008)
24. Sakamoto, R., Iwata, H.: Development of a bio feedback robot operated by simple movements to enhance the self-efficacy of dementia patients. In: The 6th International Conference on Advanced Mechatronics, pp. 308–309 (2015)

Locomotion and Telepresence in Virtual and Real Worlds

Alessandro Spada, Marco Cognetti and Alessandro De Luca

Abstract We present a system in which a human master commands in a natural way the locomotion of a humanoid slave agent in a virtual or real world. The system combines a sensorized passive locomotion platform (Cyberith Virtualizer) for the walking human, the V-REP simulation environment, an Aldebaran Nao humanoid robot with on-board vision, and a HMD (Oculus Rift) for visual feedback of the virtual or real scene. Through this bidirectional human-robot communication, the human achieves a telepresence that may be useful in different application domains. Experimental results are presented to illustrate the quality and limits of the achieved immersive experience for the user.

Keywords Telerobotic systems · Personal and entertainment robots

1 Introduction

Nowadays technology allows a convergence of different, but related domains such as telerobotics [1], wearable haptics [2] and locomotion interfaces [3], virtual and augmented reality [4, 5], and human-robot cognitive and physical interaction [6]. As a result, real or virtual objects can be tele-manipulated by haptic interfaces with force feedback, visual and tactile human perception can be enhanced by augmented virtual features, remote navigation can be realized by transferring human locomotion, and so on. While the entertainment field remains a major market offering to the public new devices to enrich multimedia experience, new application areas such as search and

A. Spada · M. Cognetti · A. De Luca (✉)
Dipartimento di Ingegneria Informatica, Automatica e Gestionale,
Sapienza University of Rome, Rome, Italy
e-mail: deluca@diag.uniroma1.it

A. Spada
e-mail: spada.aless@gmail.com

M. Cognetti
e-mail: cognetti@diag.uniroma1.it

© Springer International Publishing AG, part of Springer Nature 2019
F. Ficuciello et al. (eds.), *Human Friendly Robotics*, Springer Proceedings
in Advanced Robotics 7, https://doi.org/10.1007/978-3-319-89327-3_7

rescue, medicine, or arts and cultural e-visits are emerging, in which the human user can benefit from interacting with a remote scene by seeing, feeling, manipulating, and moving freely inside of it.

Current research developments aim at providing the best *user immersion* properties, i.e., the perception of being present in a non-physical world, independently from the user senses and stimuli [7]. We focus here on the two complementary aspects of visual feedback to the user from a real scene or from its rendering, and of the transfer of motion commands from a human to a remote (robotic) agent. These aspects represent in fact the two opposite flows of information in a human-in-the-loop control scheme.

Concerning sight in virtual reality, innovative solutions are spreading fast thanks to the capability of building relatively small and portable devices with large computing power and high resolution, capable of deceiving the human eye. One can distinguish between solutions that are developed around the user, which require a physical structure to create the virtual environment, e.g., the Cave Automatic Virtual Environment (CAVE), and those that are in contact/worn by the user, typically Head Mounted Displays. These may show information directly on the lenses, but if the user has to be immersed in a different scenario, the HMD needs to cover the human field of view and/or be inserted in a helmet with internal focusing lenses (and some distortion). The Oculus Rift [8] adopted in our work belongs to this class. Indeed, one can project on the HMD views from a virtual scene or (possibly, stereo) images from a real camera mounted on board of a mobile robot. In our setup, we used V-REP [9] as a 3D simulation environment, capable of reproducing accurately also the presence of virtual vision sensors.

In the other flow direction, the most intuitive and effective way to transfer motion commands to a robot is to replicate (through optimization techniques) the motion of a human user by precise mirroring [10, 11], rather than by using more conventional haptic devices, e.g., a joy stick for the direction/speed of walking and function buttons for stopping and starting [12]. To this purpose, tracking of the whole-body human motion relies on motion capture technology, e.g., the Optotrak in [13] or the Xsens in [14], possibly removing also the need of body markers as in [15].

However, the above works do not deal explicitly with robot navigation. In [16], the authors have used used a free-locomotion interface (a "caddie") for controlling a mobile robot by pushing its virtual representation displayed on a large projection screen. When the target is a humanoid robot that should explore a remote environment, like the Aldebaran Nao adopted in our laboratory, one may wish to replicate the actual locomotion of the human lower limbs. A simple solution is to extract the linear and angular velocity of a human walking on a locomotion platform that keeps the user approximately in place. Among such locomotion platforms, we can distinguish actuated treadmills (e.g., the Cyberwalk omnidirectional platform [3]) and passive, but sensorized supporting platforms, like the Cyberith Virtualizer [17] considered in our work. The main advantages of the latter are indeed the smaller size and its limited cost.

Based on the above analysis, we present here for the first time an original system in which a human master in locomotion on the passive Cyberith platform commands the navigation of a Nao humanoid robot, in a virtual world modeled by V-REP or in a real remote environment, while receiving immersive visual feedback on an Oculus Rift that closes the perception-actuation loop.

The paper is organized as follows. Section 2 introduces the components of the system setup and gives an overview of the adopted control architecture. More details on how telepresence was implemented are provided in Sect. 3. Section 4 reports experimental results during navigation tests, both in real and virtual environments. Conclusions and future work are summarized in Sect. 5.

2 System Setup

This section briefly describes the main hardware components of our system, see Fig. 1, together with their integration and communication.

2.1 Cyberith Virtualizer

Virtualizer is a passive omnidirectional treadmill of small size developed by the austrian company Cyberith. The structure in Fig. 1a consists of a flat base plate connected to three vertical pillars holding a circular structure, which is composed by an inner ring that can rotate inside an outer one and supports an harness. The ring structure can move up and down along the pillars in order to enable crouching. Finally, the platform has a vibration unit on the base plate in order to give haptic feedback to the user.

The working principle of the device is simple and at the same time complete. It combines a low friction principle of the base plate and a set of high-precision optical sensors with a special mechanical construction, resulting in a new form of

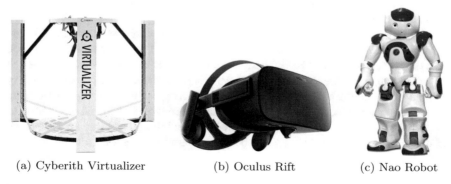

(a) Cyberith Virtualizer (b) Oculus Rift (c) Nao Robot

Fig. 1 Hardware components of the system

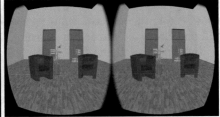

Fig. 2 Mono- (left) and stereoscopic (right) views on the Oculus Rift. The first (real) view comes from the single Nao camera, the second view from V-REP

omni-directional treadmill. The user is supposed to push the hips slightly against the ring; in this way one foot slides backwards, while the other is doing a step. The harness belt compensates the remaining friction, enabling at the same time running, walking, and crouching with user confidence and stability.

The set of API developed by the company is pretty straightforward, and consists of few calls that cover every feature. The platform returns the walker hips height and orientation, together with an average feet velocity and their direction.

2.2 Oculus Rift

Oculus Rift (see Fig. 1b) is a virtual reality headset developed by Oculus VR, composed by two Samsung AMOLED screen with a resolution of 1980×1020 pixels per eye at 90 Hz refresh rate. It has headphones and a microphone, together with a solid-state accelerometer, a gyroscope, and a magnetometer. The visual algorithm for the HMD is used to render both monoscopic and stereoscopic views (see Fig. 2). In the first case, the same camera frame is shown to both eyes, applying a suitable distortion to the image. However, the depth feeling will be lost. In the second case, the frames are taken from two (real or virtual) cameras, allowing the user to appreciate distances to the surrounding objects.

The Oculus Rift works together with a stationary infrared camera placed in the vicinity, which is delivered with the visor and has also the power and transmission unit of the device. On the HMD there is a constellation of leds installed inside the coverage, thus invisible to the human eye but visible to the infrared camera. At every sampling instant, the software takes the data and couples them with inputs from the other internal sensors. In this way, the API can return the position and orientation of the user head, relative to a fixed frame placed on the infrared camera. This sequence of measurements can be usefully mapped to a desired motion for a robot head.

2.3 Nao Humanoid Robot

In this work, the Nao humanoid developed by the french company Aldebaran was considered as the target robot —see Fig. 1c. Nao is 58 cm tall and weighs 4.3 kg. With its 25 dofs and the relatively large set of sensors (two cameras—but not as stereo pair, four microphones, sonar, IR, etc.), it is currently one of the most common robots for academic and scientific usage. Its proprietary OS enables the user to program Nao using several options, in particular setting joint position references or sending linear and angular velocity commands to the robot body, while taking care autonomously of its stabilization.

An important feature considered in the development of control modules is that all functions must be non-blocking, i.e., the robot should be able to receive multiple instructions, even interrupting the execution of the running one.

In order to virtualize the Nao robot, we adopted the multi-platform robot simulation environment V-REP [9] developed by the swiss company Coppelia. It is based on a distributed execution architecture, i.e., each object in the scene can be controlled with a script, a ROS node, a plugin and other options. Items can be mounted also on the robot, e.g., a depth sensor or a stereo camera that will produce in real time views of the virtual scene to be fed back to the Oculus.

2.4 Control Architecture

The core of the control algorithm consists of two threads working in parallel, as shown in Fig. 3. The first thread is responsible of streaming the visual output. The user has the option to stream to the HMD the images of a stereo camera mounted on the head of the robot, or those from another internal camera if not feeling comfortable with the multidimensional sensation. The programmer may also bypass the Oculus Rift to ease debugging and visualize the stereo/mono output on a screen. The first thread has also the purpose of mapping the movements of the user head to those of the robot head. The second thread is responsible instead for the robot motion, taking the velocity inputs from the Virtualizer and sending them to the robot controller. This module is responsible also for testing collisions and, in the positive case, for triggering a vibration to the platform.

A remarkable feature is that the control laws were designed to work properly both with a simulated robot and a real one, by changing only the associated IP address. In this way, it is easy to switch seamlessly the navigation from a virtual environment to the real world and vice versa.

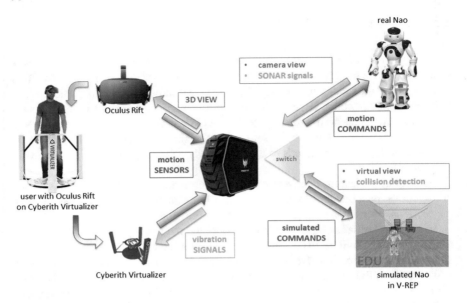

Fig. 3 Overview of the control architecture

3 Telepresence

The key to telepresence is being able to send human motion commands to the robot in a natural way and, at the same time and with little or no delay, to visualize what the robot onboard camera is currently framing, with full transparency for the user of this bidirectional communication.

As for the motion commands issued by the human, the problem can be decomposed in two subtasks, namely controlling the robot body and the robot head. To this end, when considering the local mobility of a human or of a humanoid robot, we can reduce the control problem as if both were modeled by a simple unicycle. As a result, the motion of the body for each agent can be characterized by two scalar components, a linear velocity $v \in \mathbb{R}$ and an angular velocity $\omega \in \mathbb{R}$. On the other hand, the head pose can be compactly described by a triple (α, β, γ) of roll, pitch, and yaw angles for each agent.

It is worth to remark that the velocity commands extracted from the human loco-motion on the Cyberith platform cannot be directly sent to the Nao robot, because a rescaling is needed in general. This can be done either by adjusting the dimensions of the virtual world on the fly, or by introducing a scaling factor $k_{scale} > 0$ for the transformation of commands to the robot. This scaling factor can be computed, e.g., as a simple proportion between the human and the robot maximum feasible velocities.

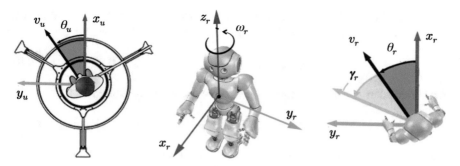

Fig. 4 Definition of axes and variables for the human and the Nao humanoid

3.1 Body Motion

With reference to Fig. 4, let the human orientation around a vertical axis be θ_u, and let v_u and ω_u be the linear and angular velocities of the human. Similarly, for the robot we have θ_r, as well as v_r and ω_r. We define the following mapping

$$v_r = k_{scale}\, v_u, \qquad k_{scale} > 0, \tag{1}$$

as the commanded linear velocity to the humanoid robot. For the angular velocity, we consider the proportional error feedback law

$$\omega_r = k_{body}\,(\theta_u - \theta_r), \qquad k_{body} > 0. \tag{2}$$

The above gain values cannot be assigned a priori, and their optimal value will depend on the specific user-robot pair. In particular, k_{scale} will take into account the maximum feasible velocity of the robot and the relative size of the virtual environment with respect to the simulated robot. In rough terms, it represents how much the robot size should be enhanced in order to approach the one of the human user. On the other hand, we have limited the angular gain to $k_{body} = 0.25$, in order to prevent overshooting during transients due to the execution delays. Moreover, larger angular gains would easily drive the robot command ω_r into saturation, since the angular rotation of the user is significantly faster than that achievable by the robot.

3.2 Head Motion

Another important aspect that affects the sensation of immersion of the user is how the human head movements are mapped to those of the robot. Indeed, the human is always able to reorient his head in any direction, while moving the body. Let γ_u be the (absolute) yaw angle of the user and γ_r the yaw angle of the robot (relative to its sagittal plane). We have considered again a proportional control law, in order to correct at each instant any relative orientation error of the robot head:

$$\omega_{head} = k_{head}\left((\gamma_u - \theta_u) - \gamma_r\right), \qquad k_{head} > 0. \tag{3}$$

When considering a discrete-time implementation of (3), we obtain

$$\gamma_{r,k+1} = \gamma_{r,k} + \omega_{head,k}\, T_c = \gamma_{r,k} + k_{head}\, T_c \left((\gamma_{u,k} - \theta_{u,k}) - \gamma_{r,k}\right) = \gamma_{u,k} - \theta_{u,k}, \tag{4}$$

where $T_c > 0$ is the sampling time, and we have chosen $k_{head} = 1/T_c$. In our experiments, the Nao robot controller accepts commands every $T_c = 10$ ms.

4 Experimental Results

Several operative conditions were tested in experiments, with the purpose of validating the strength and reliability of the control and communication code. Some tests were chosen so as to apply the proposed control laws for the body and the head separately. The following results refer to three different situations: a simple rotation, a rotation combined with a walk, and finally a complete navigation task. Each test was performed both in simulation and with experiments on a real robot, sending the output of the camera to the HMD.

4.1 Rotation

The first test is a simple rotation of the user body. As shown in Fig. 5, after a rotation of the user by 90°, the orientation error has a peak and then is exponentially corrected to zero. The robot orientation shows an initial finite delay of slightly less than 2 s and a settling time of almost 7 s. Because of this slow response, which is due to an intrinsic limitation of the mechanical locomotion of the humanoid robot, the user is supposed to wait until the robot completes its rotation. Thus, she/he needs to become confident with this temporization, adapting the own commands to the robot responsiveness. In Fig. 6a, the yaw angle of the robot head shows significant fluctuations due to the motion, whereas the signal coming from the HMD is smooth. This is translated into the noisy error signal of the relative human-robot yaw error shown in Fig. 6b.

In the experiment with the real robot we can observe a similar situation, see Figs. 7 and 8. The results from the teleoperation of the real robot are indeed perturbed by noise and affected by different delays. Although it is difficult to provide an accurate numerical estimation of the several non-idealities that influence the complete control and communication framework, these appear in general as an overall finite delay roughly in the order of 2–3 s at the body motion level, and of less than 0.5 s at the head motion level.

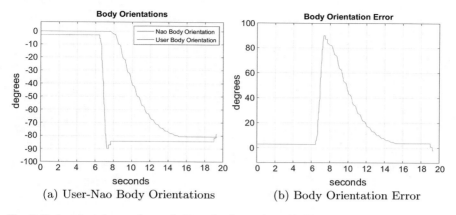

(a) User-Nao Body Orientations   (b) Body Orientation Error

Fig. 5 Body orientations and error during a simple rotation with simulated Nao

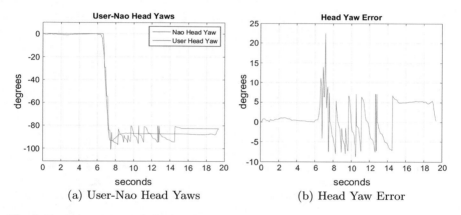

(a) User-Nao Head Yaws   (b) Head Yaw Error

Fig. 6 Head yaws and error during a simple rotation with simulated Nao

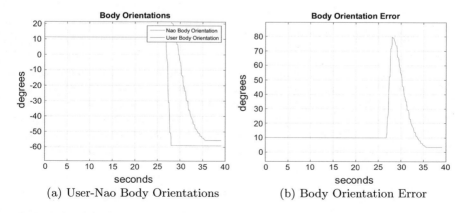

(a) User-Nao Body Orientations   (b) Body Orientation Error

Fig. 7 Body orientations and error during a simple rotation with real Nao

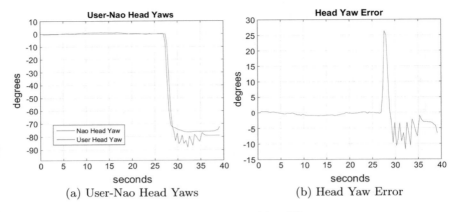

Fig. 8 Head yaws and error during a simple rotation with real Nao

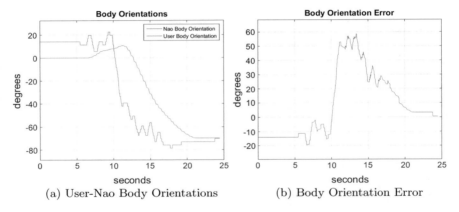

Fig. 9 Body orientations and error during a walk + rotation with simulated Nao

4.2 Rotation and Walk

The second test consists again in a rotation, combined with a walk: the user rotates the body while sliding feet on the platform. As before, the behavior of the body error in Fig. 9 is regular, with slight oscillations due to the movements of the hips during the walk. Similar observations regarding the head can be drawn from Fig. 10, as in the previous case of a simple rotation. The experimental results on body and yaw orientations and on their errors with the real Nao are shown in Figs. 11 and 12.

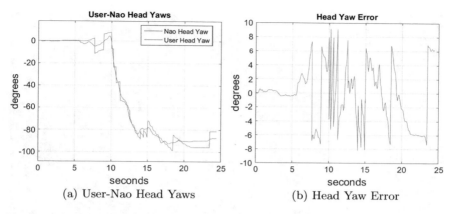

Fig. 10 Head yaws and error during a walk + rotation with simulated Nao

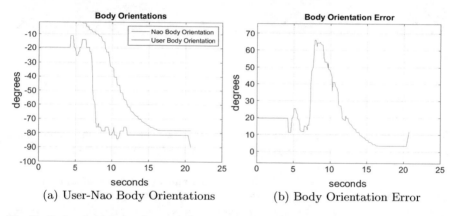

Fig. 11 Body orientations and error during a walk + rotation with real Nao

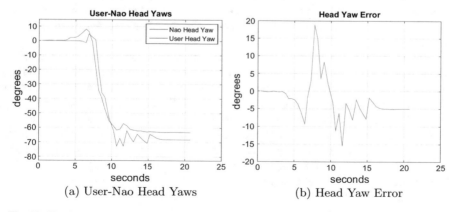

Fig. 12 Head yaws and error during a walk + rotation with real Nao

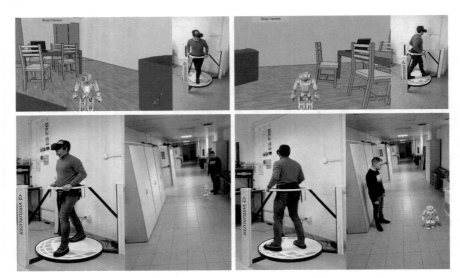

Fig. 13 Navigation in virtual [top] and real [bottom] environments

4.3 Navigation

The last experiment is a complete navigation, which can be better appreciated in the accompanying video available as https://youtu.be/RqMfbvWymFo on the YouTube channel of our laboratory.

For the simulated robot, we built a virtual room in V-REP emulating a living room. Since the robot odometry is estimated by the simulator, we observed better results as compared to those obtained with a real Nao. Also, the frame rate of the virtual camera is obviously faster and its resolution higher; thus, as expected. We observed systematically a better performance in simulated environments. Figure 13 shows some snapshots from the experiment with the simulated and with the real robot.

The results of the experiments are perturbed by several non-idealities, some independent from the robot, such as noise and delay in the communication channel, other related to the Nao itself. For instance, it is well known that this humanoid robot is affected by a significant drift in the odometry, as discussed in [18]. As time goes by, the localization error would become significant, impairing the proper execution of the programmed task. On the other hand, the success of a remotely driven real navigation will depend not only on the reliability of the control laws, but also on the quality of the visual feedback provided to the user. Thus, during longer motion tests, there will be a higher cognitive burden placed on the user in order to correct the position of the Nao and follow a trajectory to the desired goal.

5 Conclusions

We have presented a combined application of remote locomotion and human telepresence using a humanoid robot, both in a simulated virtual reality environment and in the real world. To the authors' knowledge, this is an original result in terms of realizing a complete loop between a low-cost locomotion interface, a remote humanoid in navigation, and a visual feedback to the walking user from real or virtual cameras.

The obtained results on user immersion are preliminary but encouraging. One major bottleneck was the relatively large delays present in the control loop. Indeed, the Nao robot is not equipped with a powerful processor, and due to the heavy data that the real Nao has to manage through its communication channel, the reactivity of robot body control suffers, and the frame rate of the on-board camera providing visual feedback to the user decreases significantly.

In the next future, the presented work has a large range of possible developments. First, improve the on-board vision module of Nao, by mounting on the robot a stereo camera and bring stereoscopy in the picture, as done in the simulated world. The main drawback for this robot is that the USB port on its head does not allow a good and efficient frame sending. A possible solution would be to direct the streaming onto an external router, so that the internal robot communication will only be responsible for exchanging motion data.

Second, improve the velocity mapping between the human commands and the robot by introducing a user-dependent rather than a constant scaling factor. A possible way is to learn the most appropriate value from the physiognomy of each person.

Third, remote user immersion may involve more haptics, introducing also grasping of objects. This may be implemented using a hand-tracking algorithm when the robot is not moving, e.g., through a motion sensing camera or the Oculus Touch. The first device could also be used for tracking feet, in order to climb stairs remotely.

References

1. Niemeyer, G., Preusche, C., Stramigioli, S., Lee, D.: Telerobotics. In Siciliano, B., Khatib, O. (eds.) Springer Handbook of Robotics, pp. 1085–1108. Springer (2016)
2. Scheggi, S., Meli, L., Pacchierotti, C., Prattichizzo, D.: Touch the virtual reality: using the leap motion controller for hand tracking and wearable tactile devices for immersive haptic rendering. In: Proceedings of the ACM SIGGRAPH (2015)
3. Souman, J., Robuffo Giordano, P., et al.: CyberWalk: Enabling unconstrained omnidirectional walking through virtual environments. ACM Trans. Appl. Percept. **8**(4), 24:1–24:22 (2011)
4. LaValle, S.: Virtual Reality. http://vr.cs.uiuc.edu
5. Maisto, M., Pacchierotti, C., Chinello, F., Salvietti, G., De Luca, A., Prattichizzo, D.: Evaluation of wearable haptic systems for the fingers in augmented reality applications. IEEE Trans. Haptics (2017)
6. Haddadin, S., Croft, E.: Physical human-robot interaction. In Siciliano, B., Khatib, O. (eds.) Springer Handbook of Robotics, pp. 1835–1874. Springer (2016)

7. Biocca, F., Delaney, B.: Immersive virtual reality technology. In: Biocca, F., Levy, M. (eds.) Communication in the Age of Virtual Reality, pp. 15–32. Lawrence Erlbaum, Hillsdale, NJ (1995)
8. Rendering to the Oculus Rift. https://goo.gl/WFAu1D
9. Rohmer, E., Singh, S. PN, Freese, M.: V-REP: a versatile and scalable robot simulation framework. In: Proceedings of the IEEE/RSJ International Conference on Intelligent Robots and Systems, pp. 1321–1326 (2013)
10. Suleiman, W., Yoshida, E., Kanehiro, F., Laumond, J.P., Monin, A.: On human motion imitation by humanoid robot. Proceedings of the IEEE International Conference on Robotics and Automation, pp. 2697–2704 (2008)
11. Do, M., Azad, P., Asfour, T., Dillmann, R.: Imitation of human motion on a humanoid robot using non-linear optimization. Proceedings of the 8th IEEE International Conference on Humanoid Robots, pp. 545–552 (2008)
12. Kim, T., Kim, E., Kim, J.W.: Development of a humanoid walking command system using a wireless haptic controller, pp. 1178–1183. In: Proceedings of the International Conference on Control, Automation and Systems (2008)
13. Naksuk, N., Lee, CS G., Rietdyk, S.: Whole-body human-to-humanoid motion transfer. In: Proceedings of the 5th IEEE International Conference on Humanoid Robots, pp. 104–109 (2005)
14. Koenemann, J., Burget, F., Bennewitz, M.: Real-time imitation of human whole-body motions by humanoids. Proceedings of the IEEE International Conference on Robotics and Automation, pp. 2806–2812 (2014)
15. Dariush, B., Gienger, M., et al.: Online transfer of human motion to humanoids. Int. J. Hum. Robot. 6(2), 265–289 (2009)
16. Lemoine, P., Thalmann, D., Gutiérrez, M., Vexo, F.: The "Caddie Paradigm": a free-locomotion interface for teleoperation. In: Workshop on Modelling and Motion Capture Techniques for Virtual Environments (CAPTECH), pp. 20–25 (2004)
17. Cakmak, T., Hager, H.: Cyberith virtualizer—a locomotion device for virtual reality. ACM SIGGRAPH (2014)
18. Ferro, M., Paolillo, A., Cherubini, A., Vendittelli, M.: Omnidirectional humanoid navigation in cluttered environments based on optical flow information. In: Proceedings of the 16th IEEE International Conference on Humanoid Robots, pp. 75–80 (2016)
19. Project Avatar: A Gesture-Controlled Fully Immersive Telepresence Robotics System with NAO*. https://goo.gl/oIQT8t
20. V-REP remote API. https://goo.gl/VZ3b0L

Social Robots as Psychometric Tools for Cognitive Assessment: A Pilot Test

Simone Varrasi, Santo Di Nuovo, Daniela Conti and Alessandro Di Nuovo

Abstract Recent research demonstrated the benefits of employing robots as therapeutic assistants and caregivers, but very little is known on the use of robots as a tool for psychological assessment. Socially capable robots can provide many advantages to diagnostic practice: engage people, guarantee standardized administration and assessor neutrality, perform automatic recording of subject behaviors for further analysis by practitioners. In this paper, we present a pilot study on testing people's cognitive functioning via social interaction with a humanoid robot. To this end, we programmed a social robot to administer a psychometric tool for detecting Mild Cognitive Impairment, a risk factor for dementia, implementing the first prototype of robotic assistant for mass screening of elderly population. Finally, we present a pilot test of the robotic procedure with healthy adults that show promising results of the robotic test, also compared to its traditional paper version.

Keywords Assistive robotics · Social robot · Cognitive assessment
Human-robot interaction · Dementia screening

1 Introduction

The use of robots as therapeutic assistants and caregivers is one of the most investigated application of robotics in clinical and health psychology. Indeed, the efficacy of artificial agents has been documented with children [1, 2] as well as with elderly people in a variety of neurological and psychiatric conditions [3, 4], leading scholars to suggest a stable integration of Human-Robot Interaction (HRI) in healthcare [5]. Socially Assistive Robotics is the field where most of the research has been focusing since its definition was provided [6], but service robotics and smart environments

S. Varrasi · S. Di Nuovo
Department of Educational Sciences, University of Catania, Catania, Italy

D. Conti · A. Di Nuovo (✉)
Sheffield Robotics, Sheffield Hallam University, Sheffield, UK
e-mail: a.dinuovo@shu.ac.uk

© Springer International Publishing AG, part of Springer Nature 2019
F. Ficuciello et al. (eds.), *Human Friendly Robotics*, Springer Proceedings in Advanced Robotics 7, https://doi.org/10.1007/978-3-319-89327-3_8

have been extensively explored too, especially for the elderly who may need complex assistive technology to support healthy ageing [7, 8]. Robotics can also assist with the cognitive rehabilitation [9] and to build models of cognitive dysfunctions [10].

It is evident that robots play a promising role in mental health field. However, there are still many unknown or poorly understood fields of application. One of these is the psychological assessment.

According to Scassellati [11], a great hope for robotics in Autistic Spectrum Disorder (ASD) is also the implementation of objective measurement of social behavior. This idea is encouraged by the fact ASD manifests behaviorally and diagnosis comes from the observation of developmental history and social skills, and clinicians do not always agree when evaluating the same patient [12]. Despite these solid motivations, few prototypes of ASD robotic evaluations are described in the scientific literature. Petric [13] has tried to structure a "robotic autism spectrum disorder diagnostic protocol", in order to evaluate child's reaction when called by name, his/her symbolic and functional imitative behavior, his/her joint attention, and his/her ability to communicate via multiple channels simultaneously, but the results are not clear and definitive [14]. Another diagnostic method for ASD is proposed by Wijayasinghe et al. [15], who see in HRI a way to objectively evaluate imitation deficits. In this case, the robot Zeno performs upper body gestures and the child should imitate them, while the robot automatically assesses the child's behavior.

However, it seems not much work has been done regarding other pathologies. Kojima et al. [16], for instance, published some speech recognition experiments with elderly people using the robot PaPeRo, aimed at the development of a computerized cognitive assessment system.

Therefore, at the best of our knowledge, robotic psychological assessment is almost unexplored. The evidence is limited, the pathologies studied are very few, and comparative studies (robotic assessment vs. human assessment; robotic assessment vs. computerized assessment) are not available. It is evident that more research and experimental evaluation are strongly needed.

From our point of view, the advantages of using a robotic assessor would be multiple: time-saving, quick and easy updates, widely available tools, standardization, the avoidance of assessor bias, the possibility of micro-longitudinal evaluations, scoring objectivity, and having a recording of the administration. Robots can be programmed to perform specific actions always in the same way, so standardization is one of their most interesting features. Therefore, the robotic implementation of quick screening tests could be promising, because they are often repetitive and easy to take, but time-consuming for staff. A robot could administer them, automatically score them, and transmit the result to the psychologist, who could then decide whether to continue with other human-performed tests or not. In fact, robots must not formulate diagnoses, but would provide preliminary diagnostic information about patients to reduce the workload for humans and increase the population that can be screened.

To make a step toward this direction, this paper presents results of a pilot study whose aim was to implement a screening tool for Mild Cognitive Impairment (MCI) [17] on a social robot. MCI used to be considered merely as a prodromal stage of dementia, but today it is recognized as a risk factor to develop more severe cognitive deterioration [18, 19]. It is very important to detect the so-called predictors of conversion and to determine if a patient may develop dementia for efficient planning of a prompt intervention with adequate treatment. For early detection, the markers to be considered can be both biological and psychometric [20]. However, the role of psychometric tests is crucial, because they are quicker and inexpensive, indeed cognitive deficits are usually diagnosed first this way.

This pilot study presented here had three main goals: (1) to develop a first robotic version of an MCI-specific test, fully administered and scored by a social robot; (2) to compare robotic test administration to traditional paper administration; (3) to collect data and information for further improvements.

2 Materials and Methods

2.1 Participants

As the research was not meant to provide a first validation on a clinical sample, but rather a preliminary proof of the viability of the robotic psychometric approach, we chose to enroll healthy adults ($n = 16$, Males $= 10$, Females $= 6$, M-age $= 31.5$ years, range $= 19$–61, $SD = 14.15$) among university staff and students. We selected people who had lived and worked in the UK for at least four months ($M = 110.19$, $SD = 188.23$) and we recorded their years of education as well ($M = 19.5$, $SD = 4.07$).

2.2 The MoCA Test

The Montreal Cognitive Assessment, better known as the MoCA test, is a brief cognitive screening tool for Mild Cognitive Impairment [21] freely available from the official website, used in 100 countries around the world and translated into 46 languages. It is composed of eight subtests: visuospatial/executive (alternating trail making, copying a cube, drawing of the clock), naming, memory, attention (digit span, vigilance, serial 7 s), language (sentence repetition, fluency), abstraction, delayed recall, and orientation. The maximum score is 30, and a score equal to 26 or above is considered normal. If the person has twelve years of education or less, one point is added. In this project, the Full 7.1 English version inspired the implementation, leading to a new robotic screening test for MCI. The English version of the MoCA test has two alternative versions, 7.2 and 7.3, equivalent to the main Full 7.1 version [22]. The 7.2 version was used in this study for comparison.

2.3   The Pepper Robot and the Cognitive Test Software Implementation

Our robotic cognitive test assesses the same areas of the MoCAs, therefore we have the same subtests that for simplicity have the same names.

The platform used in our experiments is the humanoid Pepper, the latest Aldebaran's and SoftBank Robotics' commercial product specifically designed for HRI and equipped with state-of-the-art interactive interfaces: touchscreen, human-like movement, pressure sensors, object recognition, speech production and comprehension, age, gender, emotion and face detector. Pepper supports different programming languages, such as Python, Java, Silverlight and C++ SDK.

To implement the prototype used in this work we used the Choregraphe suite (version 2.5.5), that provides a drag and drop interface that allows building algorithms and the robot's behaviors starting from pre-existing boxes with accessible Python code, which was modified to suit the needs of the particular implementation used in this work.

2.4   Experimental Procedure

The administration instructions reported in the English MoCA manual inspired the implementation, and two more tasks, one at the beginning of the test (the welcome task) and one at the end (the thank you task) were added. In the welcome task, Pepper introduced itself and asked the participant to provide his/her age, gender and years of education. This was meant both to collect important information about the person, as well as to train him/her on HRI. The robot could recognize and follow the face in front of it for to better engaging the participant in the interaction [23], and it moved its arms and hands as suggested in the literature in the case of HRI with adults [24].

The final result was a new psychometric test fully administered and scored by Pepper. The language of the administration was English, and the voice used was the robotic one already available in the tool. The administration was standardized, so Pepper always performed the same way, and did not change according to the participant's reactions, and it repeated the instructions only when allowed to do so by the manual. The timing of the administration was regulated by internal timers that were set empirically. Therefore, if the participant did not complete a task, the session continued when those internal timers expired. Pepper audio-recorded the whole session and took photos of the second and third tasks' drawings; moreover, it produced a Dialog file with the transcription of the verbal conversation with the participant and a Log file containing information about any technical failures that occurred, the automatic score achieved, any wrong answers received and any tasks ignored. This way, a clinical psychologist could fully review the administration and re-evaluate it if needed.

Fig. 1 Example of the human-robot interaction during the robotic administration

The administration phase was divided into two sessions for each participant: the robotic administration and the traditional paper administration. The participants were invited to enter one by one and asked, first of all, to read and sign the forms regarding the processing of personal data.

The robotic session (Fig. 1) was entirely run by Pepper: it gave the instructions, registered the answers and calculated the scoring. The experimenter did not interfere with the interaction and maintained a marginal position. The session was video-recorded and timed. After, each participant provided feedback and comments about the experience.

The traditional paper session was entirely run by the experimenter and timed as well. It is important to note that we balanced the order of the administrations by creating two subgroups: one experienced the robotic administration first, and then the traditional paper administration, while the other experienced the reverse order. Moreover, since we wanted to avoid any learning effect, for each participant the two sessions were spaced by at least five days, and the 7.2 alternative validated version of the MoCA test was used in the traditional paper session.

2.5 Data Analysis

Data collected in our pilot experiment were analyzed via classical statistical methods performed with the SPSS software (version 24).

For each participant in our experiment, we derived three scores using different evaluation modalities: (1) Standard score, which is the result of the paper and pencil administration of the test calculated using the MoCA's manual; (2) Automatic score, which is the result of the electronic test administered and automatically calculated by the robot using the build in software; and (3) Supervised score, which is the score calculated by a psychologist, who corrected the automatic score via the video and audio analysis. In our vision, modalities 2 and 3 are a simulation of the procedure for the actual application of a robotic assessment. The automatic score will be used for large population screening, while the supervised score will be calculated by a psychologist for a subset of subjects who are indicated by the automatic scoring as below the threshold and in need of a deeper analysis.

The result section presents the descriptive statistics of the three scores modalities for each subtest and the global score: minimum (Min), maximum (Max), mean (M) and standard deviation (SD). The correlations among the global scores and the subtest scores (Spearman or Pearson according to the nature of the data and the shape of the distribution) are evaluated to analyze if robot administration scores can fit the standard score. This aims to assess the concurrent external validity of the robotic procedure by comparing the automatic score with an external criterion, i.e. the MoCA score. Spearman-Brown Coefficient and Cronbach alpha are also calculated to analyze the reliability of automatic scoring, compared with the other modalities, and a regression analysis identifies a predictive function that can relate the automatic score to the standard score and allows deriving the latter from the former.

3 Experimental Results

3.1 Global Analysis

The results report that mean global automatic score is 12.69 (Min $= 6.00$; Max $= 23.00$; $SD = 4.61$), mean global supervised score is 18.62 (Min $= 10.00$; Max $= 27.00$; $SD = 4.83$) and mean global standard score is 25.00 (Min $= 21.00$; Max $= 28.00$; $SD = 2.07$), as shown in Fig. 2.

Fig. 2 Mean global scores

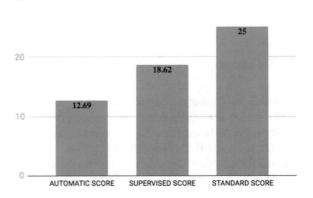

3.2   Subtest Analysis

The automatic score. In the automatic score version, some subtests do not even reach half the maximum achievable score: visuospatial/executive (0.25/5), language (0.63/3) and attention (1.31/6). The other are quite acceptable: naming $= 1.94/3$, abstraction $= 1.06/2$, delayed recall $= 2.69/5$, orientation $= 4.44/6$ (Fig. 3).

The supervised score. In the supervised score version, the subtests that do not reach half the maximum achievable score are abstraction (0.50/2), language (0.75/3) and visuospatial/executive (2.13/5). The other are as follows: naming $= 2.81/3$, attention $= 4/6$, delayed recall $= 3.13/5$, orientation $= 5.31/6$ (Fig. 4).

The standard score. In the standard score version, all the subtests reach at least half the maximum achievable score (visuospatial/executive $= 4.5/5$, nam-

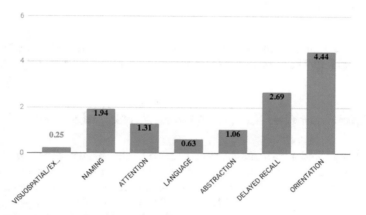

Fig. 3   Subtests of the automatic score

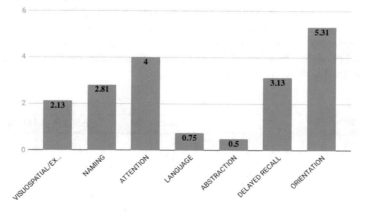

Fig. 4   Subtests of the supervised score

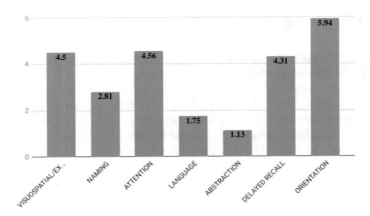

Fig. 5 Subtests of the standard score

ing = 2.81/3, attention = 4.56/6, delayed recall = 4.31/5, orientation = 5.94/6), but abstraction (1.13/2) and language (1.75/3) are lower than the others (Fig. 5).

3.3 Spearman Correlations Between the Global Scores

Spearman rank-order correlations, more suitable for the shape of distribution preliminarily assessed for these data, were calculated between the global scores (Table 1). There is a strong relationship between the supervised score and the standard score ($\rho = 0.64$), it is statistically significant ($p < 0.01$) and its strength is over the high effect-size (>0.50) according to Cohen's [25] criteria. This confirms that the robotic procedure has a promising validity. However, the correlation between the automatic and standard scores is not significant both from a statistical point of view, as well as from an effect-size point of view ($\rho = 0.01$). The correlation between the automatic and the supervised scores ($\rho = 0.38$) can be considered an indicator of inter-rater reliability and confirms, from an effect-size point of view, the fact that the supervised score is a fixed version of the automatic score.

Table 1 Spearman correlations between the global scores

Score	Automatic	Supervised	Standard
Automatic	1		
Supervised	0.38	1	
Standard	0.01	**0.64***	1

*$p < 0.01$

3.4 Pearson Correlations Between the Subtests

Pearson correlations between the corresponding subtests of the three scoring versions were calculated. In the following tables, we used these abbreviations for the subtest names: *V/E* for visuospatial/executive, *Nam* for naming, *Att* for attention, *Lan* for language, *Abst* for abstraction, *D/R* for delayed recall and *Orie* for orientation.

Automatic score subtests versus supervised score subtests. Majority automatically-scored subtests strongly and significantly correlate with the corresponding subtests of the supervised version, as shown in Table 2. It is interesting to note that automatically calculated language score also correlates with abstraction ($r = 0.67, p < 0.01$) and delayed recall ($r = 0.62, p < 0.01$) calculated in the supervised manner. This suggests that language may be involved in conceptual categorization and encoding/retrieval processes, providing an explanation for the low mean scores of the abstraction and language subtests in the standard version. If confirmed, this finding would justify the standard version's low mean global score as a problem of sampling, considering that participants were predominantly non-native speakers. The other correlations were not significant and below the medium effect-size.

Automatic score subtests versus standard score subtests. The only significant and strong correlation is between abstraction and language ($r = 0.61, p < 0.01$), confirming the finding discussed above. The other correlations are below the medium effect-size (except naming: $r = 0.32$) and not significant, as shown in Table 3.

Supervised score subtests versus standard score subtests. The attention subtest is the only one that shows a significant correlation with its corresponding subtest in the other version ($r = 0.69, p < 0.01$), while the others are not significant and below the medium effect-size, except for delayed recall versus delayed recall ($r = 0.31$). The orientation subtest, instead, correlates strongly and significantly with naming ($r = 0.63, p < 0.01$) and delayed recall ($r = 0.69, p < 0.01$), as shown in Table 4.

Table 2 Automatic versus supervised subtest correlations

		Supervised score						
		V/E	NAM	ATT	LAN	ABST	D/R	ORIE
Automatic score	V/E	0.61**						
	NAM		0.14					
	ATT			0.53*				
	LAN				0.76**	0.67**	0.62**	
	ABST					0.25		
	D/R						0.73**	
	ORIE							0.18

$*p < 0.05; **p < 0.01$

Table 3 Automatic versus standard subtest correlations (**p <0.01)

		Standard score						
		V/E	NAM	ATT	LAN	ABST	D/R	ORIE
Automatic score	V/E	0.20						
	NAM		0.32					
	ATT			0.29				
	LAN				0.28			
	ABST				**0.61**	−0.21		
	D/R						0.02	
	ORIE							−0.14

Table 4 Supervised versus standard subtest correlations (**p <0.01)

		Standard score						
		V/E	NAM	ATT	LAN	ABST	D/R	ORIE
Supervised score	V/E	0.20						
	NAM		0.18					
	ATT			**0.69**				
	LAN				0			
	ABST					−0.13		
	D/R						0.31	
	ORIE		**0.63**				**0.69**	0.06

3.5 Reliability of the Automatic Score

In order to check the reliability of the automatic score, the Spearman-Brown Coefficient and the Alpha Coefficient were calculated and found to be 0.73 and 0.67, respectively. According to the Spearman-Brown Coefficient, the automatic score shows a moderate split-half reliability. The Alpha Coefficient, then, shows an internal consistency just under the acceptable minimum (alpha < 0.70). Considering the strong limitations of the automatic scoring system that will be further discussed, it is quite surprising to find such high-reliability scores. We can justify them with the single items' mean scores, which were homogeneously low.

3.6 Multiple Linear Regression

Multiple linear regression was performed in order to find the subtests that affected the automatic and standard scores the most. The model explains most of the dependent variable's variance (Multiple R = 0.99; Multiple R-Squared = 0.98). The attention

Table 5 Subtests' influence on the variance of the automatic and standard scores

Automatic score			Standard score		
Effect	Std. coeff.	p	Effect	Std. coeff.	p
Attention	0.40	<0.01	Attention	0.79	<0.01
D/R	0.27	<0.01	D/R	0.52	<0.01
Abstraction	0.26	<0.01	Language	0.41	<0.01
Language	0.21	<0.01	V/E	0.35	<0.01
Orientation	0.20	<0.01	Abstraction	0.35	<0.01
Naming	0.19	<0.01	Naming	0.20	<0.01
V/E	0.10	<0.05	Orientation	0.12	<0.01

and delayed recall subtests affect both the automatic score and the standard score (Table 5) the most. The other independent variables significantly influence the two scores as well, but with a different percentage of variance explained.

4 Discussion

The issues occurred during the robotic administration were both automatic scoring system errors and HRI errors. They will be briefly noted, and the number of participants affected by them will be indicated in brackets.

The automatic scoring system did not assign the point if the participant repeated the digit span sequence slowly (9/16), even if it was correct; it also assigned only 0 or 3 points in the serial 7 s task, because it only recognized the five correct subtractions, while the manual allows for assigning intermediate points in case of errors (2/16). In the visuospatial/executive task, Pepper had to recognize the drawn cube and clocks with its object recognition function, but there was a high probability that the object recognition function failed, because more samples of cubes and clocks were needed to teach Pepper all the possible correct versions of drawings (16/16). Moreover, in the drawing of the clock it was possible to assign only 0 or 3 points, because the object recognition function was not sensitive enough to separately evaluate contour, numbers and hands (16/16).

Then, HRI was affected by errors as well, such as unclear pronunciation of the robot, crashes of the interfaces, low usability, lack of intuitiveness, and so on. This happened, for instance, in alternating trail making (crash of tablet: 10/16), vigilance (unclear pronunciation of the instructions: 10/16), copy of the cube (not enough time to draw: 2/16; positioning of the sheet in front of wrong sensors: 2/16; no positioning of the sheet at all: 2/16) and drawing of the clock (unclear pronunciation of the instructions: 6/16; not enough time to draw: 1/16; positioning of the sheet in front of the wrong sensors: 1/16).

Finally, it is important to note that even the standard version of the test was affected by errors, in particular, the tasks of language and abstraction showed lower mean scores than others. This finding can be explained by cultural bias because most of the participants were not native speakers. Moreover, it is important to underline that language and cultural issues may have affected all the administration, by complicating the comprehension of each task.

However, the robotic administration shows some positive aspects too. The welcome, naming, memory, fluency, abstraction and delayed recall tasks worked well, and require little and quick adjustments. Then, participants reacted positively to Pepper and they judged it as "friendly" and "cute". According to one participant, after some initial diffidence, the robot turned out to be better than PC and tablets, because HRI was "more dynamic" and "more engaging". The external concurrent validity of robotic administration is promising, even if this comes out from the correlation between the supervised score and the standard score only. This means that automatic scoring errors are the first that have to be fixed in order to obtain a better validity of the procedure, which will be further improved by the correction of HRI issues too. Even reliability is interesting and nearly acceptable, and multiple linear regression shows a good initial fit between the robotic and the standard version of the test.

5   Conclusion and Future Work

In this paper, we presented a pilot study on the use of a robotic platform for the administration and support of the scoring of a MoCA-inspired psychometric test, which is widely used for the diagnosis of Mild Cognitive Impairment. The size of the sample is small, but overall results are promising and represent a first step towards the in-depth comprehension of artificial agents' contribution to psychological assessment, even if there are many aspects that should be further investigated.

First of all, automatic scoring is prone to errors, because of the current limitation of the technology and the HRI interfaces that require many refinement cycles before being fully reliable. For example, we found that Pepper's voice should be improved to make it clearer and less childish, for instance by using a recorded human voice. Then, answer modality should allow redundant and interchangeable multimodal interfaces, so that the person can choose how to interact (e.g. speaking or touching) in case of both personal preference and technical issue.

The automatic scoring should be adequately verified and tested before actual use of any psychometric instrument. Furthermore, error causes should be investigated in detail in order to program the robotic system to flag the case for further investigation by a qualified human psychologist.

Therefore, in future work we aim to perform a test with a larger and balanced sample of native speakers, in order to perform a more accurate comparison and to investigate the effect of age and other variables on HRI. Finally, the validation on a clinical sample will be the following step if the validity and usefulness are confirmed.

Apart from the future work discussed above, there are other themes opened by this pilot study. For instance, can robots influence diagnoses? Do they affect the perception of setting and psychological assessment? Furthermore, one of the common psychometric problems—particularly in forensic settings—is the management of deception and malingering. The patient may deceive the robot and bias the test results, for example by writing down the words that s/he should recall, and the robot may not be able to notice this. The solution should come from the use of robots in controlled environments only and from video and audio recordings of the administration. However, this may lead to concerns about privacy and other ethical issues, which need to be examined in more detail.

Even if this field is completely new, we think that a robotic aid in the first phase of diagnostic path would be useful, not only to detect those that already needs clinical assistance but also to provide automatic large screening exams for prevention.

Acknowledgements The authors gratefully thank all university staff and students who participated in this study. The work was supported by the European Union's H2020 research and innovation program under the MSCA-Individual Fellowship grant agreement no. 703489.

References

1. Conti, D., Di Nuovo, S., Trubia, G., Buono, S., Di Nuovo, A.: Use of robotics to stimulate imitation in children with autism spectrum disorder: a pilot study in a clinical setting. In: Proceedings of the 24th IEEE International Symposium on Robot and Human Interactive Communication, ROMAN, pp. 1–6 (2015)
2. Conti, D., Di Nuovo, S., Buono, S., Di Nuovo, A.: Robots in education and care of children with developmental disabilities: a study on acceptance by experienced and future professionals. Int. J. Soc. Robot. **9**, 51–62 (2017)
3. Rabbitt, S.M., Kazdin, A.E., Scassellati, B.: Integrating socially assistive robotics into mental healthcare interventions: applications and recommendations for expanded use. Clin. Psychol. Rev. **35**, 35–46 (2015)
4. Conti, D., Cattani, A., Di Nuovo, S., Di Nuovo, A.: A cross-cultural study of acceptance and use of robotics by future psychology practitioners. In: Proceedings of the 24th IEEE International Symposium on Robot and Human Interactive Communication, ROMAN, pp. 555–560 (2015)
5. Iroju, O., Ojerinde, O., Ikono, R.: State of the art: a study of human-robot interaction in healthcare. Int. J. Inf. Eng. Electron. Bus. **9–3**, 43–55 (2017)
6. Feil-Seifer, D., Matarić, M.J.: Defining socially assistive robotics. In: Proceedings of the 2005 IEEE 9th International Conference on Rehabilitation Robotics, pp. 465–468 (2005)
7. Di Nuovo, A., Broz, F., Cavallo, F., Dario, P.: New frontiers of service robotics for active and healthy ageing. Int. J. Soc. Robot. **8**, 353–354 (2016)
8. Di Nuovo, A., Broz, F., Wang, N., Belpaeme, T., Cangelosi, A., Jones, R., Esposito, R., Cavallo, F., Dario, P.: The multi-modal interface of robot-era multi-robot services tailored for the elderly. Intell. Serv., Robot (2018)
9. Matarić, M.J., Scassellati, B.: Socially Assistive Robotics. In: Siciliano, B., Khatib, O. (eds.) Springer Handbook of Robotics, pp. 1973–1994. Springer International Publishing, Cham (2016)
10. Conti, D., Di Nuovo, S., Cangelosi, A., Di Nuovo, A.: Lateral specialization in unilateral spatial neglect: a cognitive robotics model. Cogn. Process. **17**, 321–328 (2016)
11. Scassellati, B.: How social robots will help us to diagnose, treat, and understand autism. In: Robotics Research, pp. 552–563 (2007)
12. Scassellati, B., Admoni, H., Matarić, M.: Robots for use in autism research. Annu. Rev. Biomed. Eng. **14**, 275–294 (2012)

13. Petric, F.: Robotic Autism Spectrum Disorder Diagnostic Protocol: Basis for Cognitive and Interactive Robotic Systems
14. Petric, F., Miklic, D., Kovacic, Z.: Robot-assisted autism spectrum disorder diagnostics using POMDPs. In: Proceedings of the Companion 2017 ACM/IEEE International Conference Human-Robot Interaction—HRI '17, pp. 369–370 (2017)
15. Wijayasinghe, I.B., Ranatunga, I., Balakrishnan, N., Bugnariu, N., Popa, D.O.: Human-robot gesture analysis for objective assessment of autism spectrum disorder. Int. J. Soc. Robot. **8**, 695–707 (2016)
16. Kojima, H., Takaeda, K., Nihel, M., Sadohara, K., Ohnaka, S., Inoue, T.: Acquisition and evaluation of a human-robot elderly spoken dialog corpus for developing computerized cognitive assessment systems. J. Acoust. Soc. Am. **140**, 2963–2963 (2016)
17. Petersen, R.C.: Mild cognitive impairment as a diagnostic entity. J. Intern. Med., 183–194 (2004)
18. Luis, C., Loewenstein, D., Acevedo, A., Barker, W.W., Duara, R.: Mild cognitive impairment: directions for future research. Neurology **61**, 438–444 (2003)
19. Landau, S.M., Harvey, D., Madison, C.M., Reiman, E.M., Foster, N.L., Aisen, P.S., Petersen, R.C., Shaw, L.M., Trojanowski, J.Q., Jack, C.R., Weiner, M.W., Jagust, W.J.: Comparing predictors of conversion and decline in mild cognitive impairment. Neurology **75**, 230–238 (2010)
20. Caraci, F., Castellano, S., Salomone, S., Drago, F., Bosco, P., Di Nuovo, S.: Searching for disease-modifying drugs in AD: can we combine neuropsychological tools with biological markers? CNS Neurol. Disord. Drug Targets **13**, 173–186 (2014)
21. Nasreddine, Z.S., Phillips, N.A., Bédirian, V., Charbonneau, S., Whitehead, V., Collin, I., Cummings, J.L., Chertkow, H.: The Montreal cognitive assessment, MoCA: a brief screening tool for mild cognitive impairment. J. Am. Geriatr. Soc. **53**, 695–699 (2005)
22. Chertkow, H., Nasreddine, Z.S., Johns, E., Phillips, N.A., McHenry, C.: The Montreal cognitive assessment (MoCA): validation of alternate forms and new recommendations for education corrections. Alzheimer's Dement. **7**, S157 (2011)
23. Xu, T. (Linger), Zhang, H., Yu, C.: See you see me: the role of eye contact in multimodal human-robot interaction. ACM Trans. Interact. Intell. Syst. **6**, 1–22 (2016)
24. Sciutti, A., Rea, F., Sandini, G.: When you are young, (robot's) looks matter. Developmental changes in the desired properties of a robot friend. In: IEEE RO-MAN 2014—23rd IEEE International Symposium on Robot and Human Interactive Communication: Human-Robot Co-Existence: Adaptive Interfaces and Systems for Daily Life, Therapy, Assistance and Socially Engaging Interactions, pp. 567–573 (2014)
25. Cohen, J.: Statistical power analysis for the behavioral sciences. Erlbaum, Hillsdale (1988)

Part III
Human Robot Collaboration

Development of a Wearable Device for Sign Language Translation

Francesco Pezzuoli, Dario Corona, Maria Letizia Corradini
and Andrea Cristofaro

Abstract A wearable device for sign language translation, called Talking Hands, is presented. It is composed by a custom data glove, which is designed to optimize the data acquisition, and a smartphone application, which offers user personalizations. Although Talking Hands can not translate a whole sign language, it offers an effective communication to deaf and mute people with everyone through a scenario-based translation. The different challenges of a gesture recognition system have been overcame with simple solutions, since the main goal of this work is an user-based product.

Keywords Sign language translation · LIS · Deaf · Data-glove · Gesture recognition

1 Introduction

There are many limitations in gesture recognition systems for sign languages translation. The first challenging tasks is the collection of movements data. A sign language is primarily expressed through the hands, but even facial expressions and full-body movements have their important meanings, so that no one of these aspects has to be neglected to achieve a perfect translation. On the other hand, the data acquisition of the user movement has be simple enough to realize a portable system.

F. Pezzuoli · D. Corona (✉) · M. L. Corradini · A. Cristofaro
School of Science and Technology, Unicam, Via Madonna delle Carceri 9,
Camerino, Italy
e-mail: dario.corona@limix.it

F. Pezzuoli
e-mail: francesco.pezzuoli@limix.it

M. L. Corradini
e-mail: letizia.corradini@unicam.it

A. Cristofaro
e-mail: andrea.cristofaro@unicam.it

© Springer International Publishing AG, part of Springer Nature 2019
F. Ficuciello et al. (eds.), *Human Friendly Robotics*, Springer Proceedings
in Advanced Robotics 7, https://doi.org/10.1007/978-3-319-89327-3_9

Second, translation of a meaningful gesture must be conducted in real time. The translation within a very large set of signs, such as an entire sign language, needs an heavy computation that can only be achieved in real time with powerful hardware and software systems. Last, a systems for sign language translation has to reconstruct the grammar structure of the phrases, because the sign languages are very different from their respective spoken languages. Different papers and studies of last decades face with these challenging tasks. For example, [1] and [2] are two surveys about the major challenges and tools for gesture recognition, in particular for video-based systems, while [3] presents the issues for an automatic sign language analysis. In [4] a framework for recognizing American Sign Language (ASL) from 3D data is presented, with a extended analysis of the sign language modeling.

In spite of these studies, nowadays there is not a commercial sign language translation system that could improve the deaf-hearing interaction.

In literature there are systems that use cameras for data acquisition (e.g. [5–8]). The major drawback of these solutions is system portability: the user is forced to be stand in front of the camera and to rearrange the camera location each time he moves from a place to another. For this reason, other systems are designed for gesture data acquisition, such as the data glove. A data-glove is a glove with sensors capable to give informations about position, velocity, acceleration and orientation of hand, arm, shoulder and fingers or some of them. In [9], different approaches for glove-based gesture recognition are presented. The data glove application for the sign language translation has a serious disadvantage: the lack of data for facial expression. Hence a complete translation of a sign language can not be achieved.

This paper describes the development of *Talking Hands*. Talking Hands is a wearable device for sign language translation that is oriented to the deaf-hearing support in ordinary life. Even if it has some limitations and it can not achieve a complete translation of the sign language, it is an user-friendly device with a great portability that allows to deaf people a basic interaction with everyone. This goals are achieved through software and design solutions that allow to simplify the different tasks.

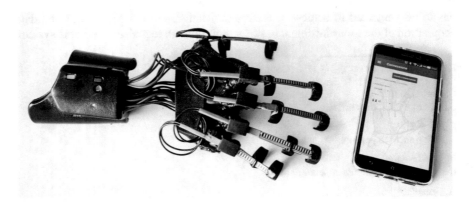

Fig. 1 Talking Hands set

The entire set of the first prototype of Talking Hands is shown in Fig. 1. The hardware of Talking Hands is composed by one data glove, that is connected with a ordinary smartphone through bluetooth. The sensors inside the data glove can measure the movements of the fingers, the hand and of the forearm. These data are processed by the microprocessor inside Talking Hands, that sends informations through bluetooth to a smartphone. A smartphone application translates the gestures and talks through a speech synthesizer.

A similar translation system, called SignSpeak, is presented in [10]. There are several advantages of Talking Hands with respect to SignSpeak in terms of hardware, software and product design:

- a customized board has been developed to reduce the dimensions and optimize the performances;
- two IMUs are used for a more accurate detection of hand movements and orientation;
- a scenario-based translation approach has been realized, that allows to achieve a high-level of translation between a large set of signs;
- the innovative design of Talking Hands does not compromise the tactile sensibility of the user.

In Sect. 2 we describe the hardware of Talking hands, so the sensors used to acquire information about the user's movements and the others electronic components. In Sect. 3 we illustrate an high level flow chart of the main parts of the translation algorithm, which is the core of the system. In Sect. 4 we describe the design, that has to overcome the issues of a wearable system for the hand. In Sect. 5 we present the result of the tests conducted on the first prototype of Talking Hands.

2 Hardware

Talking Hands does not use a commercial data-glove, but a dedicated glove is realized with a simple kind of architecture to obtain a low cost device which can be accessible by everybody. The hardware can be divided into two main modules:

- Hand Module: 10 flex sensor to detect fingers position; one IMU to detect hand orientation; one button to initialize the system.
- Arm Module: one custom board with a 32-bit microprocessor and one IMU to detect arm orientation; a led RGB to check the system status; mini-usb connection; a battery and a charge module; Bluetooth module.

A customized board has been developed as a prototyping system for the application. The board's dimensions are 3.5 mm × 5 mm. The initialization of the system through a button is required to know the initial position of the user, more precisely to get the initial data orientation of the IMUs. Once the system knows the initial position, the user can start signing and the system will correctly translate the signs.

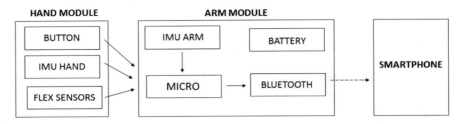

Fig. 2 Simple architecture of the system

In Fig. 2 more details about the components in the schema are presented:

- A Cortex M3 Atmel SAM3X8E is used as micro controller;
- 10 Sepctra Symbol Flex Sensors are used in the actual prototype;
- Two BOSCH BNO055 IMUs are used to detect the orientation;
- A Microchip RN42 as Bluetooth.
- A 3.7 V 1100 mAh Li-Ion Battery is used, which ensures an effective duration of 5.94 h.

A solution with 10 flex sensors has been developed to acquire data of finger bent. Two sensors are used in each finger to have the information about metacarpo and proximal inter phalangeals joints. The distal inter phalangeals joints are ignored. This solution can be improved using a flex sensor which have 2 sensible parts. In some studies (e.g. [11]) the position of the fingers are detected using EMG sensors. Even if this can be a elegant solution with only one sensor, it can not recognize the positions of single fingers, so it is not enough for a sign language translation application.

Two 9 DOF IMUs BOSH BNO055 are used to obtain information about the forearm and hand orientation. These IMUs require an initial calibration, which it last about 30 s, but they achieve a high-level of reliability even for prolonged use, thanks to the built-in functions of sensor fusion. The bluetooth is a simple BT module to communicate with the smartphone.

3 Software

Most of the works in gesture recognition use advanced mathematical tools, such as Neural Networks [12], Hidden Markov Models [4, 8], Support Vectors Machines [11], Fuzzy C-Means Clustering [13]. Talking Hands uses a simpler solution and the gesture recognition is based on a deterministic approach, using a distance function defined on the space of the sensors data. Nevertheless, this approach reaches a high level of translation, both on recognition rate and on the number of signs point of view, to some extent better than the other modeling solutions. The proposed solution can not handle with dynamic gestures and this is the major drawback. However, a satisfying communication experience can be offered to the final user.

The software of Talking Hands is composed by two main modules. The firmware pre-processes the sensors data and establishes if the user is performing a sign, i.e. a meaningful gesture. The smartphone receives data from the glove and uses the speech synthesizer to talk. The translation of the sign into a text word can be implemented both on the firmware and the smartphone application, depending on the product version.

3.1 Translation Through Scenarios

Due to the lack of facial expressions data and of a heavy computational power, the hardware of Talking Hands can translate only a limited subset of signs. In spite of these important drawbacks, Talking Hands can guarantee to a deaf person a good communication through a scenario translation, that is one of the most important novelty of this work.

We define a *Scenario* a set of signs that Talking Hands can translate in a single session. Hence the system can translate the signs of a scenario at time, that can be selected through the smartphone application. The user can switch among the scenarios on-line, i.e. during the usage without the need of re-initializing.

This approach leads to some important advantages. Across the world there are many sign languages, like the spoken ones, and some of them have different dialects, for example in the Italian Sign Language. For this reason, it is almost infeasible to realize a pre-build universal system, i.e. a system that can recognize all the sign languages. In a scenario approach, the signs can be easily recorded by the user through the smartphone application. The user can associate a sign to a word, a letter, a sound or to an entire phrase and then the sign is assigned to one or more scenarios, as shown Fig. 3. This approach enlarges the set of signs that the system can translate, without losing reliability. Hence, the same gesture can have more than one translation in different scenarios. Moreover, similar signs would not be misunderstood if they are not in the same scenarios. Since the number of possible scenarios is limited only by the memory of the smartphone, the user can have a huge set of signs, where the limitation is due mainly to the cognitive load of the user which has to remember and use properly the scenarios and their signs. In the actual prototype, the maximum number of signs in each scenario is about 40–50, but this limitation is due only to the correlations that occurs among large set of signs. Thanks to these advantages, the translation through scenarios offers a good communication for the deaf person.

3.2 Distance Function

The algorithm uses a distance on the space of vectors that represent the gestures both to establish if the user is performing a sign and to link the sign to its translation.

Fig. 3 Schematic
representation of three
simple scenarios

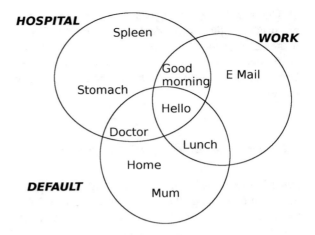

This distance is defined to be both accurate and robust, i.e. if the data-glove has one
or two broken sensors, the translation still works.

We formally introduce a proper notation to give a clear presentation of the distance
function. We use quaternions to have information about orientations of the hand and
of the forearm, with respect to the initial position, using the results in [14]. This choice
avoids the well-known gimbal-lock problem of the Euler angles and uses less bytes
than the Direct Cosine Matrix (DCM). The data coming from sensors at each sample
time are composed by 18 integers: 4 for each quaternion of the two IMUs and one
for each flex sensor. We denote the quaternions coming from the IMUs of hand and
forarm with $h = (h^1, h^2, h^3, h^4)$ and $a = (a^1, a^2, a^3, a^4)$ respectively. According to
the BOSCH BNO055 data-sheet [15], the values of h^i and a^i are in $[-2^{14}, 2^{14}]$, where
2^{14} is a scale factor. We remark that, since we are using quaternions that describe
rotations in the 3D space, the following identity holds

$$\|h\| = \sqrt{(h^1)^2 + (h^2)^2 + (h^3)^2 + (h^4)^2} = 2^{14} \tag{1}$$

so we define $Q = \{h \in \mathbb{Z}^4 : \|h\| = 2^{14}\}$. The flex sensors data are denoted with
$f = (f^1, \ldots, f^{10})$ and their values are in [0, 1000]: if $f^i = 0$, the respective finger
joint is totally bent. So we define $\mathcal{F} = [0, 1000]^{10}$.

A data package coming from sensors during a single loop of the micro controller
is denoted with

$$s = (h, a, f) \in Q^2 \times \mathcal{F} = \mathcal{S} \tag{2}$$

which is a 18 dimensional vector. With this notation, a distance function is a function

$$d : \mathcal{S} \times \mathcal{S} \to \mathbb{Z}_{\geq 0} \tag{3}$$

and it has to be equal to zero if and only if s_1 and s_2 describe the same gesture. The straightforward definition of euclidean distance is meaningless in $\mathcal{S}$, due to the two quaternion $\boldsymbol{h}, \boldsymbol{a}$ components. Hence, the two quaternions $\boldsymbol{h} = (h^1, h^2, h^3, h^4)$ and $-\boldsymbol{h} = (-h^1, -h^2, -h^3, -h^4)$ represent the same orientation, but their euclidean distance is not equal to zero. To overcome this issue, a proper distance function in $\mathcal{Q}$ must be used. Following the results in [14], we define

$$\varphi : \mathcal{Q} \times \mathcal{Q} \to [0, 1000] \qquad \varphi(\boldsymbol{h}_1, \boldsymbol{h}_2) = int \left[\frac{2000}{\pi} \arccos \left(\frac{|\boldsymbol{h}_1 \cdot \boldsymbol{h}_2|}{2^{28}} \right) \right] \qquad (4)$$

where $|\boldsymbol{h}_1 \cdot \boldsymbol{h}_2|$ indicates the absolute value of the standard dot-product. The multiplication factor $2000/\pi$ is introduced to have the same order of magnitude between the distances of quaternion and flex sensors, while the normalization factor of 2^{28} derives from (1). The distance between two data vector $s_1 = (\boldsymbol{h}_1, \boldsymbol{a}_1, \boldsymbol{f}_1)$ and $s_2 = (\boldsymbol{h}_2, \boldsymbol{a}_2, \boldsymbol{f}_2)$ is defined as

$$d(s_1, s_2) = \varphi(\boldsymbol{h}_1, \boldsymbol{h}_2) + \varphi(\boldsymbol{a}_1, \boldsymbol{a}_2) + \sum_{i=1}^{10} |f_1^i - f_2^i|, \qquad (5)$$

which is the sum of quaternions and flex sensors distances. The overall distance computation requires: 27 sums, 10 multiplications, 10 comparisons, 2 arccos evaluations. This function is very accurate and it is equal to zero if and only if s_1 and s_2 have exactly the same orientations and flex sensors values. However, it is not fault tolerant: if there is a broken sensor, the distance function does not recognize it and the whole system fails. To overcome this issue, we introduce a threshold $M \in \mathbb{Z}_{\geq 0}$ and we consider two vectors s_1 and s_2 the same gesture if

$$d(s_1, s_2) < M \qquad (6)$$

Tuning the parameter M, we trade off accuracy and robustness of the system.

3.3   Gesture Recognition Algorithm

The gesture recognition algorithm of Talking Hands is simple, deterministic and it can run in the microprocessor of the data-glove. It is a real-time checking algorithm, i.e. it processes and checks all data coming from the different sensors continuously. Hence Talking Hands does not need any external pc. In Fig. 4 the high level flow chart of the algorithm is shown.

The most important parts of the algorithm are the gesture detection, which establishes if the user is performing a meaningful gesture, and the translation, which links the sign with the corresponding output. The high level flow chart of the gesture recognition algorithm is composed by the following steps:

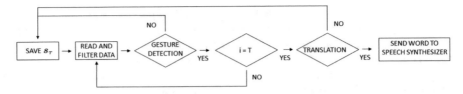

Fig. 4 High level flow chart of Talking Hands

1. Filter data to clear sensors noises;
2. Gesture Detection: determine if the user is performing a sign. To achieve this goal, the algorithm computes the distance in time of the sensors data. If the distance is larger than a given threshold, it deduces that the user is moving from a sign to another and is in a transition phase: in this case, the algorithm restarts. More precisely, the system saves the vector s_τ of the actual gestures, where τ indicates the actual loop. In the next micro controller loops, the algorithm checks if

$$d(s_\tau, s_{\tau+i}) < M \qquad \forall i = 1, \ldots, T \tag{7}$$

where $T \in \mathbb{Z}$ is the number of loops that the same gesture has to occur to be considered a sign. Consequently, T is a parameter of the system: increasing too much this value, the system would have a delay in translation; with a too low value, the system could not distinguish among signs and transition phases. If $\exists j \in 1, \ldots, T$ so that $d(s_\tau, s_{\tau+j}) > M$, the system overwrite the value of s_τ with $s_{\tau+j}$.

3. Translation: Link the sign with its meaning. A scenario can be considered a finite collection of couples (s, w), where $s \in S$ is the characteristic gesture of the sign and w_i is the string of its translation. Hence we define a scenario as $S = \{(s_i, w_i)\}_{i=1,\ldots,N}$, where N is the number of signs in the scenario S.

    - if the actual gesture s_τ is sufficient close to the characteristic vector of a recorded sign, i.e. $\exists i = 1, \ldots, N : d(s_\tau, s_i) < M$, the algorithm associates s_τ with the translation of the closer recorded sign, i.e. w_j with $j = \arg\min_{i=1,\ldots,N} d(s_\tau, s_i)$.
    - otherwise the algorithm restarts, since no translation is achieved;

4. The achieved translation w_j is sent to the speech synthesizer.

The point 2 and 3 of the previous description can be also switched: firstly the actual data are translated to the closer recorded sign if the distance is lower than a certain bound; then, if the same translation is maintained for a certain time, it is sent to the speech synthesizer. This last solution is computational expansive since it requires the comparison of each sampled data with the whole set of signs. Moreover, if the translation is performed in the smartphone application, this requires to send all the sensors data to the smartphone with a continuous communication. Hence, determining if the user is performing a sign before the effective translation is preferred. This allows to compute only a distance between the actual sensors data and the past ones in each micro controller loop. If the distance is lower than a certain bound M for a certain time T, only the last data are used for the translation, so only a sample of sensors data is sent to the smartphone, reducing the communication load of the bluetooth module.

4 Design

One of the biggest challenges for the development of Talking Hands was the design realization. The hand is the part of the human body with most articulations, so the realization of a comfortable wearable system with integrated sensors lead to many issues. Talking Hands has to guarantee the maximum portability and comfort. Moreover, it has to protect the correct operation of the hardware, especially of the sensors.

We considerate many variables, regarding physical constraints of hardware, number of sensors, positioning of the same, dimensions and arrangement of the connection wires and regarding the functionality of use in terms of wearability, transpiration, comfort. The wearable system is created to give to all sensors a life cycle as long as possible. This system is composed by:

- A set of rings for housing the different flex sensors (Fig. 5a);
- The back of the hand, including all cable connections and IMU (Fig. 5b);

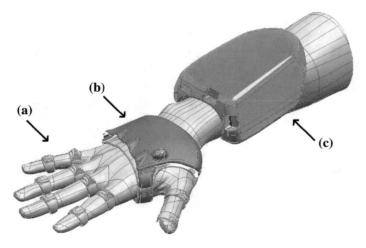

Fig. 5 Cad of Talking Hands. **a** System of rings **b** Dorso part **c** Arm part

- The forearm: connected to the previous via a fabric strap, through which the cables pass, it has the space required to insert the microprocessor, an IMU, the bluetooth communication module, and the battery (Fig. 5c).

All these parts were designed including row volumes of the hardware and wiring. A remarkable innovation of Talking Hands is the use of wearable rings that have the spaces for housing the flex sensors. This increases the reliability of flex sensors data. Another important feature is the rigid structures to insert the two IMUs in, so that the IMUs do not move during the usage and do not lose the reference system.

Every product can be custom-made to ensure the maximum comfort. We have created a parametric CAD model of Talking Hands (Fig. 5), so that we can reproduce every piece on size of the user: a 3D printer produces the external wearable module of Talking Hands after a high-quality scan of the user's hand.

5   Tests

The first prototype of Talking Hands has been carefully tested.

The translation tests are executed as follows. In each test the user performed the entire set of 40 signs for five times, for a total of 200 translations. The percentage of successful translations is shown in Fig. 6. The prototype achieves more than 90% of accuracy. The fails are both lack of translations and word misunderstanding. The second test reported in Fig. 6 achieved a translation rate of almost 80%, but with a broken sensor: this demonstrates the robustness of the system. Talking Hands has an operating time of 6 h with a 3.7 V 1100 mAh Li-Ion Battery.

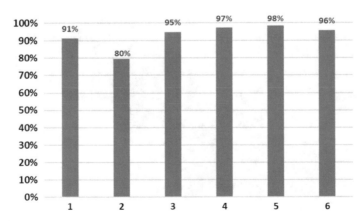

Fig. 6 Successful translation histogram

6 Future Development

The final product will achieve higher performances, thanks to more accurate flex sensors, an improved design and new hardware and software solutions.

The inclusion of dynamic gestures translation is the main software progression. The authors are already working on it with the goal of implementing a simple solution that does not compromise the final user experience. The hardware and design improvements will derive from the industrialization process. Flexible printed circuits (FPC) will minimize the row volumes of the hardware and wiring, especially for the back of the hand. Five custom flex sensors with two sensible parts will be used to improve fingers mobility. Moreover, the next-future possibilities of electronic boards, such as internet connection and machine-learning libraries, can lead to further improvements of Talking Hands.

Acknowledgements This work is supported by Limix S.r.l. (www.limix.it), an Italian start-up and spin-off of the University of Camerino. The intellectual property of Talking Hands and its different parts (hardware, software, design) is of Limix S.r.l.

References

1. Khan, R.Z., Ibraheem, N.A.: Hand gesture recognition: a literature review. Int. J. Artif. Intell. Appl. **3**(4) (2012)
2. Mitra, S., Acharya, T.: Gesture recognition: a survey. IEEE Trans. Syst. Man Cybern. Part C: Appl. Rev. **37**(3) (2007)
3. Ong, S.C.W., Ranganath, S.: Automatic sign language analysis: a survey and the future beyond lexical meaning. IEEE Trans. Pattern Anal. Mach. Intell. **27**(6) (2005)
4. Vogler, C.: American sign language recognition: reducing the complexity of the task with phoneme-based modeling and parallel hidden Markov models. Ph.D. thesis, University of Pennsylvania (2003)
5. Cooper, H., Pugeault, N., Bowden, R.: Reading the signs: a video based sign dictionary. IEEE International Conference on Computer Vision Workshops (ICCV Workshops), Nov 2011
6. Starner, T., Weaver, J., Pentland, A.: Real time American Sign Language recognition using desk and wearable computer based video. IEEE Trans. Pattern Anal. Mach. Intell. **20**(12) (1998)
7. Kelly, D., McDonald, J., Markham, C.: A person independent system for recognition of hand postures used in sign language. Pattern Recognit. Lett. **31**, 1359–1368 (2010)
8. Ho-Sub, Y., Jung, S., Younglae, J.B., Hyun, S.Y.: A person independent system for recognition of hand postures used in sign language. Pattern Recognit. **34**, 1491–1501 (2001)
9. Parvini, F., McLeod, D., Shahabi, C., Navai, B., Zali, B., Ghandeharizadeh, S.: An Approach to Glove-Based Gesture Recognition, California 90089-0781 (2009)
10. Bukhari, J., Rehman, M., Malik, S.I., Kamboh, A.M., Salman, A.: American Sign Language translation through sensory glove; SignSpeak. Int. J. u- and e-Serv. Sci. Technol. **8**(1), 131–142 (2015)
11. Akhmadeev, K.: A testing system for a real-time gesture classification using surface EMG. In: Preprints of the 20th World Congress, The International Federation of Automatic Control, Toulouse, France, 9–14 July 2017
12. Kouichi, M., Hitomi, T.: Gesture recognition using recurrent neural networks. In: ACM Conference on Human Factors in computing Systems: Reaching Through Technology (1999)

13. Xingyan, L.: Gesture Recognition Based on Fuzzy C-Means Clustering Algorithm. Departement of Computer Science, The University of Tennessee Knoxville
14. Huynh, D.Q.: Metrics for 3D rotations: comparison and analysis. J. Math. Imaging Vis. **35**, 155–164 (2009)
15. Sensortec, B.: BNO055 Intelligent 9-axis absolute orientation sensor. https://cdn-shop.adafruit.com/datasheets/BST_BNO055_DS000_12.pdf

Close Encounters of the Fifth Kind? Affective Impact of Speed and Distance of a Collaborative Industrial Robot on Humans

Marijke Bergman and Maikel van Zandbeek

Abstract So called "close encounters of the fifth kind" between robots and humans will become more frequent as the use collaborative robots is spreading. The purpose of the current study is to obtain a better understanding of how humans experience the behavior of an industrial robot arm moving around in their vicinity while mounted on a work surface. Such experiences may parallel those that occur in human-human interaction. This study focusses on the effects of speed and stopping distance in a live confrontation of participants (N = 90) with the robot arm and on obtaining affective responses from these humans directly and extensively. The results indicate that both speed and stopping distance of an industrial robot arm have significant effects on the perceived safety and affective state of the participants. Consequences for design are discussed. To ensure optimal collaboration between human and robot, robot design should make use of various communicative cues. Research on, for instance, non-verbal communication between humans, may provide useful insights.

Keywords Perceived safety · Affective responses · Subjective experience Collaborative robot · Human-robot interaction

1 Introduction

The current, so called fourth technological revolution will change the way we work with and operate machines and robots. Robots develop into collaborative robots, becoming automated colleagues instead of isolated machines. What are known as "close encounters of the fifth kind", bilateral contact experiences through cooperative communication [1], will become more frequent. In this context it is important to ensure optimal teamwork and interaction between humans and robots. Research on

M. Bergman (✉) · M. van Zandbeek
Research Group People and Technology, Institute of Human Resources
Management and Psychology, Fontys University of Applied Sciences,
Eindhoven, The Netherlands
e-mail: m.bergman@fontys.nl

© Springer International Publishing AG, part of Springer Nature 2019
F. Ficuciello et al. (eds.), *Human Friendly Robotics*, Springer Proceedings
in Advanced Robotics 7, https://doi.org/10.1007/978-3-319-89327-3_10

127

human teams indicates that interpersonal trust plays an important role in optimal collaboration and team performance [2]. Trust has been identified as an important element for the success of human-robot teamwork as well [3]. Humans should be able to trust that a collaborative robot does not harm their interests and welfare [4]. When they, operators in particular, do not feel comfortable or safe around a collaborative robot, for instance because it comes too close or moves too fast, this may result in avoidance and stress [5].

For industrial robots revised safety standards (e.g. ISO/TS 15066:2016) are supplemented by standards for collaborative industrial robot systems and the work environment [6]. The main focus of current standards and related research, is on ensuring that a collaborative robot is intrinsically safe and will not physically harm its human operator or humans in its vicinity [7–10]. Thus there is a rise in attention for designing robots that operate safely and a need to understand what is required to ensure that humans will not be injured [11].

However this approach does not fully take into account the subjective safety as perceived by humans. Yet such perceived safety may be considered as differing from objective safety [12]. Perceived safety concerns the perception of risk and danger by the human interacting with the robot as well as the experienced comfort [13]. These will be compromised if the robot and its behavior do not enable humans to experience positive emotions or affect in its vicinity. Also, not feeling safe and experiencing negative affect and emotions such as fear and tension, may result in automatic behavioral responses like fight or flight reactions. Fight or flight are natural physiological reactions in humans and other mammals to perceived danger or threat that rapidly activate the body to confront or escape this threat [14]. Perceived danger can, of course, be caused by actions of human colleagues, but may also be the result of actions of machines and robots. In industrial settings this might cause accidents.

Consequently the subjective experiences and affective state of humans working with or around collaborative robots should be investigated and used as input to ensure seamless interaction, communication and collaboration between robots and humans. It is particularly important that the design of a collaborative robot is well tuned to humans and their needs. The acceptability and use of robots and other automated systems are influenced by, amongst others, trust, workload and perceived safety [12, 15]. The most influential factors to build trust are robot-related rather than human- or environment-related [4]. Specifically performance factors including the behavior, reliability and predictability of the robot, and robot attributes, such as proximity and (assumed) personality, are found to be important.

Much work on Human Robot Interaction (HRI) and human friendly design concerns social robots interacting in public space, [amongst others 16–18]. Insights in human friendly design and perceived safety of industrial collaborative robots are important, but rather underexposed [13]. Still some important factors, that come into play to ensure perceived safety and positive affect, emerge from studies on both social and industrial robots. Robot-related factors that are repeatedly found to influence affective state, perceived safety and trust, are speed, distance and predictability or legibility.

Both the relative *speed* of movement and of approach of a robot have a demonstrable effect on affective state, perceived safety and comfort. Several studies show that a higher speed of about 1 m/s is rated as uncomfortable [19–21] as compared to lower speeds between 0.254 and 0.381 m/s. There may however be a tradeoff between experienced safety and satisfaction with working speed [22]. Such findings run parallel to knowledge about interaction between animals as well as between humans. Animals are known to display fear and flight reactions that are correlated with speed, size and directness of approach of other animals [23].

Perceived safety or comfort is compromised not only by higher speed but also by the *distance* of the robot's reach [22] or proximity [24]. The optimal distance seems to depend on the appearance of the robot [20], its size, and the task or context it is used for. Intermediate distances (1.22–2.44 m) within the so called social zone are generally experienced as comfortable [21, 24, 25]. It may thus be assumed that humans apply conventions regarding social space when they interact with robots in a similar way as when interacting with other humans.

The affective state and task-performance of humans is further influenced by the *predictability* of robot motions: lower predictability results in lower performance [19] and lower experienced comfort [20, 25]. Levels of perceived safety and trust may depend on whether a robot, either social or industrial, meets the expectations of humans [24, 26]. In continuation of this, in human-human interaction people tend to attribute meaning to posture and movement of others [27]. Further, humans are inclined to attribute life to non-living objects (animism) and to interpret non-human beings and objects and their behavior in human terms (anthropomorphism) [28]. Thus knowledge about the interpretation of non-verbal behavior in human-human interaction could be projected onto robot-human interaction as well.

For industrial collaborative robots, in particular robot arms, the evidence on the effects of speed and distance on the affective state of humans in their vicinity, is rather indirect. Some experiments use simulations to investigate such effects [19], some studies derive suspected affective state from physiological measures or the behavior of the humans involved [25], others use one single question on comfort, or use different measures, such as workload [21]. The purpose of the current study is to obtain a better understanding of how humans experience the behavior of an industrial robot arm mounted on a work surface while moving around in the their vicinity. This study focusses on the effects of speed and stopping distance in a live confrontation of humans with the robot arm and on obtaining affective responses from these humans directly and extensively. This results in the following research question:

- *What is the effect of speed and stopping distance of an industrial robot arm on the affective state and perceived safety of humans?*

Based on previous research we expect both higher speed and shorter stopping distances to result in lower scores on perceived safety and positive affect and higher scores on negative affect.

2   Method

2.1   *Participants*

Two experiments were set up to test both the effects of speed and of stopping distance on affective state and perceived safety. A total of 90 subjects participated in this study. Fifty participants were assigned to Experiment 1 (speed) and 40 participants were assigned to Experiment 2 (stopping distance). All participants have a background in Engineering, either as a student or as a member of the school of Engineering at Fontys University of Applied Sciences, Eindhoven, The Netherlands. This group is considered to be representative for people who may work with industrial robots, though the level of education of actual workers may be lower. Of the participants, five were female and 85 were male, their age ranging from 18 through 63 years.

2.2   *Procedure*

Participants in the two experiments were exposed to a Universal Robots UR5 type robot arm, which was mounted on a surface at table height. Figure 1 shows the actual situation in the laboratory as well as a schematic setup for the experiments. The participants were instructed to stand at a fixed spot marked on the floor and remain there while the robot arm was moving. The edge of the fixed spot was located at 50 cm from the midpoint of the closest edge of the surface. The movements of the robot arm were controlled by the researcher, who was standing at the side of the setup. The acceleration of the arm movement was 120 cm/s in both experiments. After the movement exposure the participants filled out a questionnaire on their affective responses. They did so in the same room, so they could view the set up if needed.

Experiment 1. In the first experiment (N = 50) the effect of speed on perceived safety and affective state was tested. Participants were assigned at random to either one of two conditions: a *low speed* condition (N = 25), or a *high speed* condition (N = 25). In the low speed condition the robot arm accelerates at 120 cm/s and moves at 25 cm/s. This speed is chosen on the basis of the existing standards for robot safety. In the high speed condition the robot arm accelerates at 120 cm/s and moves at 40 cm/s. The speed in this second condition is limited by the stability of the mounting of the arm. In both conditions the same movement pattern is executed three times for each participant.

Experiment 2. In the second experiment (N = 40) the effect of stopping distance of the arm on perceived safety and affective state is tested. Participants are assigned at random to either one of two conditions: a *short distance* condition (N = 20), or a *long distance* condition (N = 20). In the short distance condition the robot arm approaches at 20 cm/s and stops at 11.25 cm from the participant in average (7.5–15 cm depending

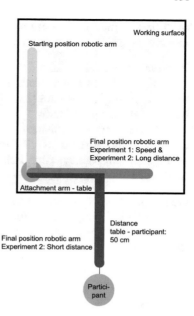

Fig. 1 Experimental setup: actual situation and schematic view for both experiments

on the abdominal extent of the participant, standing on the fixed spot at 50 cm from the edge of the surface). In the long distance condition the robot arm approaches at 20 cm/s and stops at the edge of the surface, 50 cm from the fixed spot. In both conditions the same movement pattern is executed three times for each participant.

2.3 Measures

Affective state was measured using a questionnaire consisting of five statements on perceived safety, six items on positive affect, five items on negative affect and one open question on the experience in general. A rating scale method was used as subjective measure to investigate the affective responses.

Perceived Safety. Five statements were formulated to measure perceived safety. Participants responded on a 7-point rating scale (1 = totally disagree through 7 = totally agree). The statements are derived from the questionnaire by Lasota and Shah [29], that is designed to measure perceived safety and comfort in the presence of an industrial robot. It contains statements like "The robot moved too fast to feel comfortable" or "I trusted the robot would not hurt me". In the current study the item scale is internally consistent (Cronbach's $\alpha = 0.830$).

Positive and Negative Affect. The items on positive and negative affect are derived and translated into Dutch from the categories likeability and perceived safety of the Godspeed questionnaire [13]. Six individual word items were used to measure positive affect, e.g. "friendly", "relaxed" and five items to measure negative affect, e.g. "unpleasant", and "fearful". Participants responded on a 7-point rating scale (1 = totally disagree through 7 = totally agree). The item scales used were tested on internal consistency: Cronbach's $\alpha = 0.915$ for positive affect and Cronbach's $\alpha = 0.887$ for negative affect.

Open Question. One open question was added to ensure the participants could indicate their feeling even it did not match any of the items. This question was presented at the beginning of the item list to prevent any priming from the other items.

3 Results

A one-way between groups analysis of variance (ANOVA) was used for both experiments to investigate the effect of speed respectively stopping distance on experienced safety, positive affect, and negative affect. The statistical test results reported here are the essential test statistic, F-ratio (F) together with the degrees of freedom between groups and within groups, the likelihood (p) of the size of the F-ratio occurring by chance, and the effect size (η^2) to assess the proportion of variability in the data that can be attributed to the independent variable. The assumption of homogeneity of variance is violated in both experiments, therefore both Welch and Brown-Forsyth tests for equality of means were performed.

3.1 Experiment 1: Speed

All measures, experienced safety, positive affect, and negative affect, show a significant effect of speed. The results are summarized in Table 1. For *Perceived Safety* ($F(1,48) = 9.23, p = 0.003, \eta^2 = 0.161$) participants in the low speed condition show higher perceived safety ($M = 5.63, SD = 0.84$) than participants in the high speed condition ($M = 4.76, SD = 1.16$).

For *Positive Affect* ($F(1,48) = 9.63, p = 0.003, \eta^2 = 0.167$) participants in the low speed condition show a higher score ($M = 4.69, SD = 1.03$) than participants in the high speed condition ($M = 3.54, SD = 1.15$).

Table 1 ANOVA testing effects of speed on perceived safety, positive affect and negative affect

	ANOVA	Low speed M and SD	High speed M and SD
Perceived safety	$F(1,48)=9.23$, $p=0.003$, $\eta^2=0.161$	$M=5.63$, $SD=0.84$	$M=4.76$, $SD=1.16$
Positive affect	$F(1,48)=9.63$, $p=0.003$, $\eta^2=0.167$	$M=4.69$, $SD=1.03$	$M=3.54$, $SD=1.15$
Negative affect	$F(1,48)=4.25$, $p=0.045$, $\eta^2=0.081$	$M=2.21$, $SD=0.94$	$M=2.89$, $SD=1.20$

For *Negative Affect* ($F(1,48)=4.25$, $p=0.045$, $\eta^2=0.081$) participants in the low speed condition show a lower score ($M=2.21$, $SD=0.94$) than participants in the high speed condition ($M=2.89$, $SD=1.20$).

3.2 Experiment 2: Stopping Distance

The effect of stopping distance is significant for all measures, experienced safety, positive affect, and negative affect. The results are summarized in Table 2. For *Perceived Safety* ($F(1,38)=34.31$, $p<0.001$, $\eta^2=0.474$) participants in the condition with long stopping distance show higher perceived safety ($M=5.97$, $SD=0.65$) than participants in the short stopping distance condition ($M=4.40$, $SD=1.00$).

For *Positive Affect* ($F(1,38)=24.156$, $p<0.001$, $\eta^2=0.389$) participants in the condition with long stopping distance show higher mean scores ($M=5.31$, $SD=0.61$) than participants in the short stopping distance condition ($M=3.64$, $SD=1.39$).

For *Negative Affect* ($F(1,38)=24.126$, $p<0.001$, $\eta^2=0.388$) participants in the condition with long stopping distance show lower mean scores ($M=1.91$, $SD=0.49$) than participants in the short stopping distance condition ($M=3.52$, $SD=1.38$).

Open Question. The open question revealed a variety of affective states both positive ("fun", "jolly", "fine", translated from Dutch) and negative ("shocked",

Table 2 ANOVA testing effects of stopping distance on perceived safety, positive affect and negative affect

	ANOVA	Long distance M and SD	Short distance M and SD
Perceived safety	$F(1,38)=34.31$, $p<0.001$, $\eta^2=0.474$	$M=5.97$, $SD=0.65$	$M=4.40$, $SD=1.00$
Positive affect	$F(1,38)=24.156$, $p<0.001$, $\eta^2=0.389$	$M=5.31$, $SD=0.61$	$M=3.64$, $SD=1.39$
Negative affect	$F(1,38)=24.126$, $p<0.001$, $\eta^2=0.388$	$M=1.91$, $SD=0.49$	$M=3.52$, $SD=1.38$

"exciting", "uneasy", translated from Dutch). Yet no striking differences in these responses were found in between the conditions in either experiment. It was, however, notable that several participants mentioned they would like to have more information on the movements and trajectory of the robot arm.

4 Discussion

4.1 Conclusion

The current study shows that both speed and stopping distance of an industrial robot arm have significant effects on the perceived safety, positive affect and negative affect of the participants. Higher speed of movement results in a lower perceived safety, lower scores on positive affect and higher scores on negative affect. The effect of speed on perceived safety seems rather small, yet significant. In general, it can be concluded that participants prefer the lower speed and do not feel as comfortable when the speed is higher. This outcome concurs with previous research [19–21].

Similar effects were found for stopping distance. A short stopping distance results in lower perceived safety, lower scores on positive affect and higher scores on negative affect. The differences between the mean scores for long and short stopping distances seem to be even more prominent than for speed. It seems that participants feel more comfortable with a larger distance between the robot arm and themselves. Thus it appears that negative effects of the violation of personal space and a preference for working at a distance conforming to the so called social space applies to industrial contexts as well as social contexts [21, 24, 25].

An important observation is that noticeable differences in affective responses occur, even though the participants in this study have experience with robots and industrial robot arms. This suggests that being used to working in such a setting cannot completely erase the emotional reactions of humans. They may be slightly muted and may become more or less unconscious. However they will still have an effect at a subconscious level and may cause stress, fatigue, avoidance or discomfort in the workplace [4, 5]. Since sub-optimal trust within the robot-human team can impair successful collaboration [3, 4] the effects found in this study are worth taking into consideration. What is even more important, is that subconscious reactions and reflexes may result from affective states and consequently compromise safety. Such reactions and reflexes, for instance fight or flight, are not easily handled by technical safety measures and traditional standards.

4.2 Limitations

The current study consisted of live experiments where participants rated their impressions on answer scales. Being in an experiment requires that participants are informed before it starts. Also the researcher was present and visible during the experiment. These circumstances could bias the answers on the questionnaire [30]. Secondly, self-reporting has some drawbacks. It requires a certain level of conscious awareness and the ability to verbalize affect and emotions. Thus the ratings may not be the best representation of actual feelings. The items used, are based on two existing questionnaires used to investigate affective state and perceived safety. Translating the items into Dutch and adapting or selecting items can cause some distortions in the measures. Fine-tuning of items and questionnaires is important to further develop usable and valid instruments. Furthermore, even though it may be assumed that affective state has effects on behavior, it is not fully a reliable predictor of such behavior. Additional information on the affective state, for instance the amount of stress, may be obtained by physical measures. Observational studies could give more information on actual behavior.

In this study movement repetition was limited to three times. Though the participants have ample experience with industrial robot arms and their movements, an effect of movement was still found. It would however be interesting to see if any habituation would occur when exposed to the same movements more often. The background of the participants may be comparable to the actual users of robots in an industrial environment. However, external validity may be limited since possible confounding effects of level of education, age and experience with robots were not taken into account. Further, actual users may have less technical affinity than the participants in this study. This may show itself in more negative affective reactions. They may take longer to accept and to get used to collaborative robots and their movements. Also encounters with a moving robot arm in a laboratory may be quite different to work settings in factories. In this study for instance, participants had no active role in interacting with the robot arm. These differences could have effects on the levels of perceived safety and trust in real world situations [31]. Further research with operators and other employees in industrial environments should complement the current outcomes.

4.3 Consequences for Design

The current study as well as earlier research show that speed and stopping distance of robots have significant effects on perceived safety and the affective state of humans. The design characteristics of a robot will have to ensure objective safety, but will also have to accommodate the preferences of humans encountering and working with them. Though it is becoming clear which speeds and stopping distances are optimal in general, the specific context has to be taken into account as well. For instance, though low speed is experienced as safer, humans may get irritated if the robot moves too slow to work together productively [22]. This would require options for customization of the settings by the operator, so that an optimal balance between objective and subjective safety is achieved.

Both speed and stopping distance can be seen as communicative cues that are interpreted by humans. As was pointed out in the introduction, humans tend to attribute intentions to robots and machines in general, due to animism and anthropomorphism [28]. Also they interpret movements as carrying messages, even if there is no communicative intention [27]. It may be conjectured that the movements of a robot arm will contain, probably unintentional, cues for humans as well. These cues are used to determine the intentions of the robot and will influence not only affective state, but behavior as well, for instance approaching the robot or moving away from it. To ensure future positive "encounters of the fifth kind", that is cooperative communication, the design of a robot should make use of various communicative cues. The participants in the current study indicated they would like to have more information on the movements and trajectory of the robot arm. This does not necessarily call for humanoid industrial robots, but does require grounded use of visual, audio and tactile cues, for instance lights to indicate the intended direction of movement or sound to signal approach. This way safe and seamless interaction between human and robot can be brought closer and promote true teamwork.

Acknowledgements The authors would like to thank all participants of the experiments. Additionally, Henk Kiela, Frank Schoenmakers of the research group Mechatronics and Janienke Sturm of the research group People and Technology are thanked for their support. This work was partly subsidized by the Regional Attention and Action for Knowledge Circulation scheme (RAAK) which is managed by the Foundation Innovation Alliance (Stichting Innovatie Alliantie [SIA]) with funding from the Dutch Ministry of Education, Culture and Science (OCW). SIA-RAAK had no role in the study design; the collection, analysis, and interpretation of the data.

References

1. Center for the Study of Extraterrestrial Intelligence: The CE-5 initiative. http://new.cseti.org/ce5initiative.html
2. Sheng, C.W., Tian, Y.F., Chen, M.C.: Relationships among teamwork behavior, trust, perceived team support, and team commitment. Soc. Behav. Pers. Int. J. **38**(10), 1297–1305 (2010)
3. Charalambous, G., Fletcher, S., Webb, P.: The development of a scale to evaluate trust in industrial human-robot collaboration. Int. J. Soc. Robot. **8**(2), 193–209 (2016)
4. Hancock, P.A., Billings, D.R., Schaefer, K.E., Chen, J.Y., De Visser, E.J., Parasuraman, R.: A meta-analysis of factors affecting trust in human-robot interaction. Hum. Factors **53**(5), 517–527 (2011)
5. Arai, T., Kato, R., Fujita, M.: Assessment of operator stress induced by robot collaboration in assembly. CIRP Ann. Manuf. Technol. **59**(1), 5–8 (2010)
6. Fryman, J.: Updating the industrial robot safety standard. In: 41st International Symposium on Robotics; Proceedings of ISR/Robotik, pp. 1–4 (2014)
7. Fryman, J., Matthias, B.: Safety of industrial robots: from conventional to collaborative applications. In: ROBOTIK 2012; 7th German Conference on Robotics, pp. 1–5 (2012)
8. Haddadin, S., Albu-Schäffer, A., Hirzinger, G.: Requirements for safe robots: measurements, analysis and new insights. Int. J. Robot. Res. **28**, 11–12 (2009)
9. Zanchettin, A.M., Ceriani, N.M., Rocco, P., Ding, H., Matthias, B.: Safety in human-robot collaborative manufacturing environments: metrics and control. IEEE Trans. Autom. Sci. Eng. **13**(2), 882–893 (2016)
10. Vasic, M., Billard, A.: Safety issues in human-robot interactions. In: 2013 IEEE International Conference on Robotics and Automation (ICRA), pp. 197–204 (2013)

11. Haddadin, S., Haddadin, S., Khoury, A., Rokahr, T., Parusel, S., Burgkart, R., Albu-Schäffer, A.: On making robots understand safety: embedding injury knowledge into control. Int. J. Robot. Res. **31**(13), 1578–1602 (2012)

12. Coeckelbergh, M., Pop, C., Simut, R., Peca, A., Pintea, S., David, D., Vanderborght, B.: A survey of expectations about the role of robots in robot-assisted therapy for children with ASD: ethical acceptability, trust, sociability, appearance, and attachment. Sci. Eng. Ethics **22**(1), 47–65 (2016)

13. Bartneck, C., Kulić, D., Croft, E., Zoghbi, S.: Measurement instruments for the anthropomorphism, animacy, likeability, perceived intelligence, and perceived safety of robots. Int. J. Social Robot. **1**(1), 71–81 (2009)

14. Teatero, M.L., Penney, A.M.: Fight-or-flight response. In: Milosevic, I., McCabe, R.E. (eds.) Phobias: the psychology of irrational fear, pp. 179–180. ABC-CLIO, Santa Barbara (2015)

15. Michalos, G., Makris, S., Tsarouchi, P., Guasch, T., Kontovrakis, D., Chryssolouris, G.: Design considerations for safe human-robot collaborative workplaces. Procedia CIrP **37**, 248–253 (2015)

16. Rossi, A., Dautenhahn, K., Koay, K.L., Saunders, J.: Investigating human perceptions of trust in robots for safe HRI in home environments. In: Proceedings of the Companion of the 2017 ACM/IEEE International Conference on Human-Robot Interaction, pp. 375–376 (2017)

17. Satake, S., Kanda, T., Glas, D.F., Imai, M., Ishiguro, H., Hagita, N.: How to approach humans? Strategies for social robots to initiate interaction. In: 4th ACM/IEEE International Conference on Human-Robot Interaction (HRI), pp. 109–116 (2009)

18. Sisbot, E.A., Marin-Urias, L.F., Broquere, X., Sidobre, D., Alami, R.: Synthesizing robot motions adapted to human presence. Int. J. Soc. Robot. **2**(3), 329–343 (2010)

19. Koppenborg, M., Nickel, P., Naber, B., Lungfiel, A., Huelke, M.: Effects of movement speed and predictability in human-robot collaboration. Hum. Factors Ergon. Manuf. Serv. Ind **27**(4), 197–209 (2017)

20. Butler, J.T., Agah, A.: Psychological effects of behavior patterns of a mobile personal robot. Auton. Robots **10**(2), 185–202 (2001)

21. Tan, J.T.C., Duan, F., Zhang, Y., Watanabe, K., Kato, R., Arai, T.: Human-robot collaboration in cellular manufacturing: design and development. In: IROS 2009. IEEE/RSJ International Conference on Intelligent Robots and Systems, pp. 29–34 (2009)

22. Koene, A., Remazeilles, M.P., Prada, M., Garzo, A., Puerto, M., Endo, S., Wing, A.M.: Relative importance of spatial and temporal precision for user satisfaction in human-robot object handover interactions. In: Proceedings of the New Frontiers in Human-Robot Interaction, vol. 14 (2014)

23. Stankowich, T., Blumstein, D.T.: Fear in animals: a meta-analysis and review of risk assessment. Proc. R. Soc. Lond. B Biol. Sci. **272**(1581), 2627–2634 (2005)

24. Rios-Martinez, J., Spalanzani, A., Laugier, C.: From Proxemics theory to socially-aware navigation: a survey. Int. J. Soc. Robot. **7**(2), 137–153 (2015)

25. Bortot, D., Ding, H., Antonopolous, A., Bengler, K.: Human motion behavior while interacting with an industrial robot. Work **41**(Suppl 1), 1699–1707 (2012)

26. Eder, K., Harper, C., Leonards, U.: Towards the safety of human-in-the-loop robotics: challenges and opportunities for safety assurance of robotic co-workers. In: The 23rd IEEE International Symposium on Robot and Human Interactive Communication, pp. 660–665 (2014)

27. Pollick, F.E., Paterson, H.M., Bruderlin, A., Sanford, A.J.: Perceiving affect from arm movement. Cognition **82**(2), 51–61 (2001)

28. Guthrie, S.E.: Faces in the Clouds: A New Theory of Religion. Oxford University Press, New York (1993)

29. Lasota, P.A., Shah, J.A.: Analyzing the effects of human-aware motion planning on close-proximity human-robot collaboration. Hum. Factors **57**(1), 21–33 (2015)

30. Kruse, T., Pandey, A.K., Alami, R., Kirsch, A.: Human-aware robot navigation: a survey. Robot. Auton. Syst. **61**(12), 1726–1743 (2013)

31. van den Brule, R., Dotsch, R., Bijlstra, G., Wigboldus, D.H., Haselager, P.: Do robot performance and behavioral style affect human trust? Int. J. Soc. Robot. **6**(4), 519–531 (2014)

Dynamic Graphical Signage Improves Response Time and Decreases Negative Attitudes Towards Robots in Human-Robot Co-working

Iveta Eimontaite, Ian Gwilt, David Cameron, Jonathan M. Aitken, Joe Rolph, Saeid Mokaram and James Law

Abstract Collaborative robots, or 'co-bots', are a transformational technology that bridge traditionally segregated manual and automated manufacturing processes. However, to realize its full potential, human operators need confidence in robotic co-worker technologies and their capabilities. In this experiment we investigate the impact of screen-based dynamic instructional signage on 39 participants from a manufacturing assembly line. The results provide evidence that dynamic signage helps to improve response time for the experimental group with task-relevant signage compared to the control group with no signage. Furthermore, the experimental group's negative attitudes towards robots decreased significantly with increasing accuracy on the task.

Keywords Human–robot collaboration · Dynamic graphical signage · Negative attitudes towards robots · Manufacturing · Technology acceptance

I. Eimontaite (✉) · D. Cameron · J. M. Aitken · S. Mokaram · J. Law
Sheffield Robotics, The University of Sheffield, Pam Liversidge Building,
Sir Frederick Mappin Building, Mappin Street, Sheffield S1 3JD, UK
e-mail: i.eimontaite@sheffield.ac.uk

D. Cameron
e-mail: d.s.cameron@sheffield.ac.uk

J. M. Aitken
e-mail: jonathan.aitken@sheffield.ac.uk

S. Mokaram
e-mail: s.mokaram@sheffield.ac.uk

J. Law
e-mail: j.law@sheffield.ac.uk

I. Gwilt · J. Rolph
Art & Design Research Centre, Sheffield Hallam University, Cantor Building,
153 Arundel Street, Sheffield S1 2NU, UK
e-mail: i.gwilt@shu.ac.uk

J. Rolph
e-mail: joseph.rolph@student.shu.ac.uk

© Springer International Publishing AG, part of Springer Nature 2019
F. Ficuciello et al. (eds.), *Human Friendly Robotics*, Springer Proceedings
in Advanced Robotics 7, https://doi.org/10.1007/978-3-319-89327-3_11

139

1 Introduction

The manufacturing sector is poised to undergo massive change, with Industry 4.0, the Internet of Things, and the Digital Agenda all leading toward greater digitalization and connectivity of processes. Collaborative robots, or 'co-bots', are a key driving technology that will blur the boundaries between traditional manual and automated manufacturing processes. Combining the flexibility of the human workforce with the precision and repeatability of robotics allows shared workspaces to emerge where uncaged robots and humans interact directly. The development of intuitive and natural interfaces, collaboratively with users, will lead to greater levels of human-robot interaction, allowing the operation and reconfiguration of complex robotic systems with less training and shorter setup times.

As the requirements on autonomy, complexity and safety of robots increase, human operators need to develop confidence in robotic collaborative processes and understand the capacities of the robots they are working with so that effective collaboration can occur. One of the essential requirements for this confidence to be built is an appropriate level of trust [1, 2]. Too low or too high a level of trust can lead to errors and greater task completion times [3]. Another issue is that, although robots in manufacturing are not a new phenomenon, workers can still feel threatened by their presence and perceived control of the working environment. Feeling out of control, especially in situations perceived as threatening, can also result in higher stress levels [4, 5]. Whereas understanding the requirements of unfamiliar situations, and having the necessary knowledge and information, can result in individual empowerment and a sense of control [6], as well as a decrease in stress levels [5, 7, 8]. Finally, an individual's cognitive load is often already high in manufacturing [9] and there can be little capacity beyond undertaking a complex activity for monitoring co-workers progress (human or automated). This issue is exacerbated if users feel they do not have enough information or training to undertake a task. While increased cognitive load can lead to decreased concentration on the task, performance, and increased number of accidents [10, 11], establishing effective measures can reduce the amount of information necessary for efficient decision-making [12].

From the issues discussed above, it is evident that effective information communication to aid human–robot interaction in manufacturing settings can play a positive role [13]. Information communication via graphical signage can be a viable tool in improving issues around human robot collaboration. The main merits of graphical signage are that it (i) displays clear instructions for individuals with little or no prior experience [14, 15]; (ii) does not depend on language, as opposed to written instructions, making it suitable for multicultural environments and beneficial for non-native speakers [16]; (iii) does not depend on voice control, making it suitable for noisy environments such as factories; and (iv) decreases cognitive load and requires less information processing than written instructions [17]. In the healthcare domain, information communication has been proven to be an effective way to increase human well-being; for example, providing concise and clear information increases patients' eagerness to discuss their situation and prompt questions [18] leading to feelings of being in control and able to make important decisions [6], which in turn can decrease experienced stress [5, 7, 8]. An alternative option is social cues (facial expressions, body language, pitch of voice) which have similar benefits to graphical signage [19]. However, for robots that do not having animate-like form (such as robotic arms), exhibiting social cues can become ambiguous to interpret as a form of providing information. This type of communication is perhaps better used in studies with anthropomorphic robots (such as Baxter).

The aim of the current research was to further extend the findings of our previous graphical signage investigation [20] by examining the effects of dynamic screen-based signage on collaborative human-robot interaction within a manufacturing workforce. We explored this by observing the behavior of employees from our industrial collaborators with little or no experience in working with robotics in a manufacturing context. It was expected that experimental group participants, who were presented with task relevant dynamic signage, would have higher task completion accuracy and lower response times compared with control group participants with no signage. Furthermore, we predicted that negative attitudes towards robots and robot anxiety would decrease after the experiment for both experimental and control groups, but that the decrease would be greater for the experimental group participants.

2 Methods

2.1 Participants

Forty low-skilled product assemblers (9 female) from the partner's workforce participated in the experiment across two groups (20 per group). One participant was removed from the analysis due to leaving the pre-test questionnaire empty, resulting in 39 participants for the final analysis. The mean age of participants was 38.63 (SD = 13.30). Participants had no prior knowledge of using robots and they were not exposed to the signage before the experiment. This participant population was ideal for the study, as the company was preparing to install its first collaborative robot cell, but the employees had not been trained, or interacted with a robot before. The work was approved by the University of Sheffield Ethics Committee.

2.2 KUKA iiwa Lightweight Arm

In this study, a KUKA LBR iiwa 7 R800 was used for the human–robot co-working task. The KUKA iiwa is developed as a collaborative robot, specifically allowing direct human–robot interaction, and has a set of configurable safety measures suited to co-working (Fig. 1). For this study the robot was set to be operated in a compliant

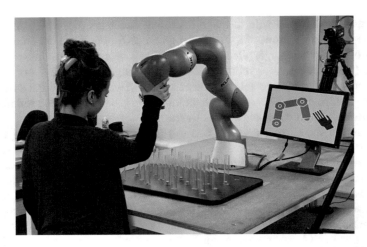

Fig. 1 KUKA iiwa and experimental task setup. On the screen graphics are displayed indicating permitted interaction

safe mode 'T1' with limits on speed and a requirement for human monitoring. The KUKA iiwa was controlled via our own Application Programming Interface (API) [21].

2.3 Design of the Graphical Signage

For the project a set of bespoke graphical symbols was refined from earlier paper-based solutions [20] and developed further into dynamic graphical signage to provide real time information to the user about robot operational processes. The signage was collaboratively designed in workshops with workers from the industry partner, before being refined. Dynamic graphical signage visually represented human-robot interaction events to provide a co-worker with key information, such as when it is safe to touch the robot, the expected speed of robot movement, and operational area, etc.

During the experiment, screen-based dynamical graphic signage was presented on the computer monitor (20 inch screen diameter) on the right side of the robot, 70 cm away from the desk edge where participants were standing. Experimental group participants were presented with animated gifs with information about the robotic arm (direction of robot movement (x and y axes), the speed and reach of robot, applied force from the user to control the robot; each presented for 30 s at the start of interaction with the robot). Being dynamic allowed the signage to communicate nuanced information relating to participants interaction with the robot. During the trials, the signage indicated when participants should position the robot over a tube,

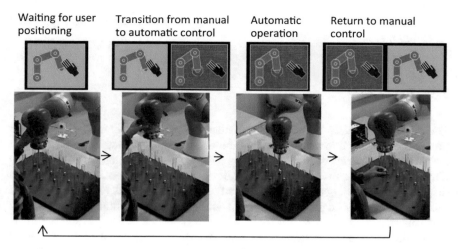

Fig. 2 Sequence of dynamic signage display relating to "touch" and "do not touch" the robot

and when robot was completing the process on its own (Fig. 2). Control group participants were presented with a blank screen during the experiment.

2.4 Measures

The following measures were used in the experiment:

Negative Attitudes towards Robots Scale (NARS). This scale consisting of 14 statements was developed by Nomura et al. [22]. The sub-scale of attitudes towards interactions with robots was administered pre- and post-experiment where participants had to indicate their agreement with each of the statements on five-point scale (1—strongly disagree, 5—strongly agree).

Robot Anxiety Scale (RAS). This scale measures anxiety affecting participants' interactions with robots [23]. The sub-scale measuring anxiety towards the behavioral characteristics of robots were conducted pre- and post-experiment. Participants had to indicate how anxious they feel about each statement on a six-point scale (from 1—I do not feel anxiety at all to 6—I feel very anxious).

Behavioral Measures. The following behavioral measures were the main interest of the study: (1) participant accuracy (collected bolts/number of trials), and (2) time taken to complete the task. These measures serve as behavioral indexes of task achievement.

To control for confounding variables, measures of participants risk taking attitudes [24], their experience with robotics [25], computer use frequency, and programming expertise were taken. Furthermore, a post experiment questionnaire asked participants to indicate which signs they have seen during the experiment (attention measure).

2.5 Procedure

The study was conducted on site at the industrial partner's factory, in one of their process development rooms in order to achieve a realistic working experience. Upon arrival at the experiment, participants signed a consent form and filled in a questionnaire measuring their demographic information and the control variables (participant's robot anxiety (RAS), negative attitude towards robots (NARS), computer use frequency, age, and programming experience, risk taking attitude and experience with robots).

The process to be undertaken by participants was described in the following way: "on the table there are 16 narrow tubes and 6 of them contain M5 bolts (Fig. 1). These bolts need to be put into a collection box, however they are inaccessible to the human (the tubes being too narrow to allow access by hand), and, although the robot can reach and pick the bolts, it is unable to locate in which tubes they are placed". Participants could only complete the task by collaborating with the robotic arm and they were not provided with any additional verbal information.

Whilst the experimental group were provided with screen-based dynamic graphical signage, the control group were presented with a blank screen. Although effective collaboration requires good communication, the task was intuitive enough to complete without additional information. In fact, all participants successfully completed at least two trials. In this particular study, the control group was used to compare the effects of signage effects versus no signage on participants' wellbeing (attitudes and anxiety towards robot) and performance (accuracy and response time).

The scenario was not taken from an existing process, but had been carefully designed to reflect anticipated future human-robot co-working interactions.

The maximum time to complete the task was 10 min. The experiment was recorded on video to obtain behavioral measures. During the experiment, a collaborator observes the participants' performance as a safety measure in case the experiment needed to be aborted.

Participants were informed that they were going to be video recorded during the experiment, and the material collected would be used for data coding and further statistical analysis. However, measures were taken to keep the data anonymous and confidential.

After the main part of the experiment, participants' robot anxiety (RAS) and negative attitudes towards robots (NARS) were measured once again. Participants had to fill in a signage effectiveness and recollection questionnaire as a control measure for their attention to signage. The whole experiment lasted around 30 min.

2.6 Analysis

The study used a mixed design with between-subject and repeated measures. It contained two independent conditions: signage relevant to the task (experimental), and no signage (control). Repeated measures within conditions were used: participant's first completed baseline measures of attitudes and anxiety towards robots, and again after the robot interaction scenario.

3 Results

3.1 Group Differences

A preliminary check using an independent t-test was run to examine pre-trial distribution of participants across two participant groups (experimental and baseline control) and showed no significant differences between experimental and control groups ($t(37) \leq 1.44, p \geq 0.159$).

As a second control measure, gaze duration towards where the signage would/would not be presented (measured in number of frames) was recorded. This control measure was taken to verify that the computer monitor was not a distractor in itself, and that that the experimental group participants were looking at the signs. Results showed that the experimental group participants had a significantly longer gaze duration compared to control group participants ($t(22.94) = -3.93, p = 0.001$). A further measure of signage recollection showed that participants had seen the signs (with 80% accuracy in indicating which signs they have seen) while control group participants indicated that they had not seen any signage.

3.2 Dynamic Graphical Signage Effects on Performance

An investigation of task completion accuracy between experimental and baseline control groups with ANOVA (dependent variable accuracy rate, independent variable—condition) showed that overall participants performance was not significantly different between the groups ($F(1, 35) = 0.45, p = 0.505$).

To investigate whether response time was affected by signage, Linear mixed models (between-subject—condition, within-subject–trial number (1–6), covariate—tube position) was performed on participants' response time on successfully completed trials. The analysis showed a significant main effect of condition ($F(1, 179) = 10.28, p = 0.002$) and main effect of trial ($F(5, 132) = 2.65, p = 0.025$) as well as significant Condition by Trial interaction ($F(5, 132) = 2.34, p = 0.045$; Fig. 3a).

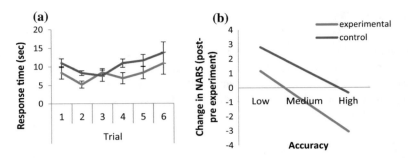

Fig. 3 **a** Response time as a function of task trial in experimental and control groups (±SEM); **b** Change in NARS score (post-pre-experiment) as a function of task accuracy modulated by participant group. Red line—control group, blue line—experimental group participants

3.3 Robot Anxiety and Negative Attitudes Towards Robots

A moderated regression with independent variable of accuracy, moderator of group (experimental and baseline control), and dependent variables of change between post- and pre-experiment RAS and NARS (two separate models), was done using PROCESS syntax.

The analysis showed that accuracy on the task predicts the change of negative attitudes towards robots moderated by condition ($F(4, 33) = 3.29$, $p = 0.0226$, $R^2 = 0.29$, Fig. 3b). The post-trial NARS score decreased compared to the pre-trial NARS score as task accuracy increased, yet this was significant only in the experimental group ($b = -11.28$, $t = -2.66$, $p = 0.0119$), but not the control group ($b = -8.43$, $t = -1.89$, $p = 0.0671$).

No other models were significant either with NARS and predictor response time, or equivalent analysis with RAS ($F(4, 33) \leq 0.341$, $p \geq 0.796$, $R^2 \leq 0.17$).

4 Discussion

This project explored the effect dynamic graphical signage has on participants' performance, negative attitudes, and robot anxiety on a manufacturing-type HRI task [26]. It was found that showing dynamic signage: reduced the response time for completing each trial of the task; decreased negative attitudes towards robots as a function of increased accuracy of low skilled manufacturing participants with no prior experience with robots.

These results cannot be explained by group differences as participants' age, computer usage for work and leisure, gaming, NARS, RAS, RTI, programming and robot experience did not significantly differ between groups.

The main finding of the study provides evidence that graphical signage decreases response time. The dynamic signage provided information for the participant about changes in the process, and therefore could help to complete each trial more quickly without needing to unnecessarily adjust the robot position (adding more time for the trial completion). An argument could be made that effective communication is needed for collaboration, and the control group did not receive this communication as they did not have signs. However, all the participants collected at least two bolts suggesting that they understood the process enough to partially complete the task while in collaboration with a robot.

Finally, the results showing improved response times are consistent with previous studies showing that with more information about a particular task, individuals' efficiency increases. For example, navigation of unfamiliar settings takes less time with signage [27, 28].

The second aim of the current study was to investigate the effects of signage on participants' negative attitudes towards robots and robot anxiety. The findings revealed a decrease in post-trial negative attitudes towards robots correlating with increasing accuracy, but this decrease was significant only for the experimental group participants. A possible explanation for this result is that participants' sense of empowerment and knowledge of the processes they are going through has increased. Graphical signage is designed to help people understand the requirements of unfamiliar situations, and this information can lead to greater empowerment and a sense of control [6], and decrease the levels of stress experienced [5, 7, 8]. Additionally, negative attitudes towards robots decrease after having interacted with robot [29], and having information about robot abilities and maneuverability might have influenced participants' expectations of robot performance [30].

The presented results are promising and raise several opportunities for further research. First of all, some signs were not considered intuitive, but although participants were neither trained on the task, nor had been exposed to similar signs in the factory, the signage still reinforced their decisions on how to operate robot. One opportunity for future development would be to refine the signs further to make them more intuitive and clear, and re-assess their impact on users. At the same time, a longitudinal study is needed to fully explore the effects of the signage; investigating the performance after participants became confident in the robotic system would provide further evidence of how dynamic signage can aid human-robot collaboration. Finally, although beyond the scope of the current study, future experiments could also look into comparing graphical signage versus voice control or text instructions. Comparing the effects of different modalities of information communication would allow determination of their drawbacks and strengths, and support the development of industrial application guidance.

To summarize, this study confirms and extends the results from our previous study by progressing from static to dynamic signage [20, 26]. Dynamic screen-based signage, which is presented at a specific time of relevancy, has been shown to decrease task completion time compared to the no-signage condition. Furthermore, the experimental group's negative attitudes towards robots significantly decreased in correlation with increasing accuracy on the task. Taken together, these results indicate that graphical signage can not only improve efficiency on the task, but also improve participants' comfort when compared to participants receiving no signage.

References

1. Cameron, D., Aitken, J.M., Collins, E.C., Boorman, L., Chua, A., Fernando, S., McAree, O., Martinez-Hernandez, U., Law, J.: Framing factors: the importance of context and the individual in understanding trust in human-robot interaction (2015)
2. Hancock, P.A., Billings, D.R., Schaefer, K.E., Chen, J.Y.C., de Visser, E.J., Parasuraman, R.: A meta-analysis of factors affecting trust in human-robot interaction. Hum. Factors J. Hum. Factors Ergon. Soc. **53**, 517–527 (2011)
3. Lee, J.D., See, K.A.: Trust in automation: designing for appropriate reliance. Hum. Factors J. Hum. Factors Ergon. Soc. **46**, 50–80 (2004)
4. Mathews, A., Mackintosh, B.: A cognitive model of selective processing in anxiety. Cogn. Ther. Res. **22**, 539–560 (1998)
5. Ozer, E.M., Bandura, A.: Mechanisms governing empowerment effects: a self-efficacy analysis. J. Pers. Soc. Psychol. **58**, 472 (1990)
6. Ussher, J., Kirsten, L., Butow, P., Sandoval, M.: What do cancer support groups provide which other supportive relationships do not? The experience of peer support groups for people with cancer. Soc. Sci. Med. **62**, 2565–2576 (2006)
7. Lautizi, M., Laschinger, H.K.S., Ravazzolo, S.: Workplace empowerment, job satisfaction and job stress among Italian mental health nurses: an exploratory study. J. Nurs. Manag. **17**, 446–452 (2009)
8. Pearson, L.C., Moomaw, W.: The relationship between teacher autonomy and stress, work satisfaction, empowerment, and professionalism. Educ. Res. Q. **29**, 37 (2005)
9. Thorvald, P., Lindblom, J.: Initial development of a cognitive load assessment tool. In: The 5th AHFE International Conference on Applied Human Factors and Ergonomics, 19–23 July 2014, Krakow, Poland, pp. 223–232. AHFE (2014)
10. Bahar, G., Masliah, M., Wolff, R., Park, P.: Desktop reference for crash reduction factors (2007)
11. Laughery, K.R.: Safety communications: warnings. Appl. Ergon. **37**, 467–478 (2006)
12. Chen, R., Wang, X., Hou, L.: Augmented reality for collaborative assembly design in manufacturing sector. Virtual Technol. Bus. Ind. Appl. Innov. Synerg. Approaches Innov. Synerg. Approaches **105** (2010)
13. Chan, A.H.S., Ng, A.W.Y.: Investigation of guessability of industrial safety signs: effects of prospective-user factors and cognitive sign features. Int. J. Ind. Ergon. **40**, 689–697 (2010)
14. Tufte, E.R.: The visual display of quantitative information. Graphics Press, Connecticut (1993)
15. Frixione, M., Lombardi, A.: Street signs and Ikea instruction sheets: pragmatics and pictorial communication. Rev. Philos. Psychol. **6**, 133–149 (2015)
16. Ben-Bassat, T., Shinar, D.: Ergonomic guidelines for traffic sign design increase sign comprehension. Hum. Factors J. Hum. Factors Ergon. Soc. **48**, 182–195 (2006)
17. Lamont, D., Kenyon, S., Lyons, G.: Dyslexia and mobility-related social exclusion: the role of travel information provision. J. Transp. Geogr. **26**, 147–157 (2013)
18. Mills, M.E., Sullivan, K.: The importance of information giving for patients newly diagnosed with cancer: a review of the literature. J. Clin. Nurs. **8**, 631–642 (1999)
19. Elprama, S., El Makrini, I., Vanderborght, B., Jacobs, A.: Acceptance of collaborative robots by factory workers: a pilot study on the role of social cues of anthropomorphic robots. In: The 25th IEEE International Symposium on Robot and Human Interactive Communication, pp. 919–919 (2016)
20. Eimontaite, I., Gwilt, I., Cameron, D., Aitken, J.M., Rolph, J., Mokaram, S., Law, J.: Assessing graphical robot aids for interactive co-working. In: Schlick, C., Trzcieliński, S. (eds.) Advances in Ergonomics of Manufacturing: Managing the Enterprise of the Future, pp. 229–239. Springer International Publishing, Cham (2016)
21. Mokaram, S., Martinez-Hernandez, U., Aitken, J.M., Eimontaite, I., Cameron, D., Law, J., Rolph, J., Gwilt, I., McAree, O.: A ROS-integrated API for the KUKA LBR iiwa collaborative robot. In: Proceedings of the 20th World Congress of the International Federation of Automatic Control (IFAC), Toulouse, France (2017)

22. Nomura, T., Suzuki, T., Kanda, T., Kato, K.: Measurement of negative attitudes toward robots. Interact. Stud. **7**, 437–454 (2006)
23. Nomura, T., Suzuki, T., Kanda, T., Kato, K.: Measurement of anxiety toward robots. In: The 15th IEEE International Symposium on Robot and Human Interactive Communication, 2006. ROMAN 2006, pp. 372–377. IEEE (2006)
24. Nicholson, N., Soane, E., Fenton-O'Creevy, M., Willman, P.: Personality and domain-specific risk taking. J. Risk Res. **8**, 157–176 (2005)
25. MacDorman, K.F., Vasudevan, S.K., Ho, C.-C.: Does Japan really have robot mania? Comparing attitudes by implicit and explicit measures. AI Soc. **23**, 485–510 (2009)
26. Eimontaite, I., Gwilt, I., Cameron, D., Aitken, J.M., Rolph, J., Mokaram, S., Law, J.: Graphical languages improve performance and reduce anxiety during human-robot co-working. PLoS ONE (under review)
27. Tang, C.-H., Wu, W.-T., Lin, C.-Y.: Using virtual reality to determine how emergency signs facilitate way-finding. Appl. Ergon. **40**, 722–730 (2009)
28. Vilar, E., Rebelo, F., Noriega, P.: Indoor human wayfinding performance using vertical and horizontal signage in virtual reality: indoor human wayfinding and virtual reality. Hum. Factors Ergon. Manuf. Serv. Ind. **24**, 601–615 (2014)
29. Stafford, R.Q., Broadbent, E., Jayawardena, C., Unger, U., Kuo, I.H., Igic, A., Wong, R., Kerse, N., Watson, C., MacDonald, B.A.: Improved robot attitudes and emotions at a retirement home after meeting a robot. In: RO-MAN, 2010 IEEE, pp. 82–87. IEEE (2010)
30. Muir, B.M.: Trust between humans and machines, and the design of decision aids. Int. J. Man-Mach. Stud. **27**, 527–539 (1987)

Towards Progressive Automation of Repetitive Tasks Through Physical Human-Robot Interaction

Fotios Dimeas, Filippos Fotiadis, Dimitrios Papageorgiou,
Antonis Sidiropoulos and Zoe Doulgeri

Abstract In this paper, a novel method is developed that enables quick and easy programming in repetitive industrial tasks, through kinesthetic demonstration from the operator. The robot learns the task cycle with the assistance of haptic cues and progressively transitions from manual into autonomous operation using a novel variable stiffness control strategy and assistive virtual fixtures. The training process, requires a small amount of iterations, decreasing dramatically the typical robotic programming time. In the experimental evaluation, an operator is able to program a pick and place task in less than a minute, without requiring any interaction with a user interface or pre-programming of the task sequence.

Keywords Physical human-robot interaction · Variable stiffness control
Virtual fixtures · Dynamic movement primitives

1 Introduction

The programming of repetitive robot tasks (e.g. pick and place) in an industrial environment is a very common but time-consuming procedure, demanding expert personnel and significant financial burden. Utilizing conventional offline

F. Dimeas (✉) · F. Fotiadis · D. Papageorgiou · A. Sidiropoulos · Z. Doulgeri
Automation & Robotics Lab, Department of Electrical & Computer Engineering,
Aristotle University of Thessaloniki, Thessaloniki, Greece
e-mail: dimeasf@ee.auth.gr

F. Fotiadis
e-mail: ffotiadis@auth.gr

D. Papageorgiou
e-mail: dimpapag@eng.auth.gr

A. Sidiropoulos
e-mail: antonis.sidiropoulos@issel.ee.auth.gr

Z. Doulgeri
e-mail: doulgeri@eng.auth.gr

© Springer International Publishing AG, part of Springer Nature 2019
F. Ficuciello et al. (eds.), *Human Friendly Robotics*, Springer Proceedings
in Advanced Robotics 7, https://doi.org/10.1007/978-3-319-89327-3_12

programming techniques, a simple task that lasts only a few seconds to execute, can take many hours to program [1]. Owing to the need of expert operators and their associated cost, robotic systems are rendered non-viable for small and medium-sized enterprises, where changes in the production line are frequent. Robot programming by demonstration (PbD) is a promising way to reduce programming time and enable such enterprises to benefit from automated production in handling and assembly. In one of the simplest forms of PbD, physical human-robot interaction is utilized as kinesthetic guidance, where the users can guide the robot with direct manipulation and demonstrate the task, usually by pre-programming of the task sequence [2] and manually selecting the via-points.

Motivated by the potentials of physical human-robot interaction, we investigate the transition between the manual guidance and the autonomous operation towards a novel way of intuitive PbD for repetitive tasks that facilitates fast and easy programming by demonstration. The main objective of this work is to develop a framework that allows an operator to demonstrate a repetitive robotic task with kinesthetic teaching and to provide active assistance to the operator while the robot learns to execute the task autonomously. To that aim, the robot has to recognize and then execute the demonstrated task, without requiring prior knowledge or pre-programming. Starting from full compliance, the robot will gradually become increasingly pro-active in the task execution, while allowing the user to make corrections and adjustments. After a few demonstrations, the task will be executed autonomously and the user can stop interacting with the robot, formulating the concept of progressive automation.

The proposed progressive automation framework utilizes a variable stiffness controller for continuous role allocation, based on the agreement level between consecutive demonstrations. While the operator demonstrates a path that is close to a previously demonstrated one and applies low forces indicating agreement, the target stiffness increases, providing the leader's role to the robot. The framework introduces virtual fixtures to assist the operator in accurately repeating the same path and thus accelerating progressive automation. When a high deviation occurs from the nominal path or a high force is applied, the stiffness decreases, revoking the leading role to the operator and allowing adjustments. The demonstrated path is encoded with Dynamic Movement Primitives (DMP) that offer compact representation with the ability to generalize in changes of the environment. The task cycle is detected using haptic cues, which are also used for segmenting the task. The contribution of this work lies in the combination of DMP for trajectory encoding with a novel variable stiffness strategy and assistive virtual fixtures towards progressive automation in less than a minute.

1.1 Related Work

Within the concept of learning from demonstration [3], Gaussian Mixture Models and DMP are widely used to learn from demonstration trajectories [4, 5] and interaction forces [6]. Focusing on handling tasks, PbD for pick and place was conducted in [7] using task primitives, which are subjected to the low accuracy of the motion capturing system and the limitation of the primitive vocabulary. Tracking of 3D motion and gesture recognition that was proposed in [8] for pick and place tasks, involves high computational cost and limited accuracy. Kinesthetic teaching for pick and place tasks was used in [9] along with a neural network that was trained to sort different objects. In this approach the operator had to provide rewards for the training and the switching to automatic mode was conducted instantly without the ability for adaptation.

The aforementioned methods include two distinctive steps, one for the demonstration and another for the execution, that are considered to hinder the wide adoption of such programming techniques. As a result, the operator must frequently interact with the robot's interface to manually switch from manual into automatic control mode and vice versa. The need for ease of use and fast deployment of robots with minimum amount of demonstrations is highlighted in various related works [10, 11]. Focusing on repetitive robot tasks in industrial environment, the approach proposed here for PbD that is characterized by progressive automation, involves continuous role allocation between the human and the robot. The main open issues that are investigated by the research community in role allocation include *how* and *when* the leader-follower roles should be exchanged for optimal interaction [12–14], particularly in effort sharing tasks. By learning force and motion patterns from human demonstrations, the robot in [15] was given cognitive capabilities for improved haptic cooperation. A continuous role adaptation technique was proposed in [16] where the operator can exchange the lead role with the robot based on the applied interaction forces. The higher this force, the more active role the operator obtains. In that way, the operator can take the lead and make adjustments to the predefined trajectory that is executed by the robot. Unlike [16], the role allocation method that we propose takes into consideration, in addition to the interaction force, the position deviation from the reference path. This allows continuous role allocation based on the agreement level between the demonstrations, with the role rate of change depending on the current stiffness value. Variable stiffness control, which is usually utilized for assisting the cooperation [17] and for improving stability [18], was also used for role allocation in [19]. In particular, the authors in [19] proposed a method to allow proactive behavior of the robot during interaction with a human, but with constant and experimentally tuned gains.

Similar to the objectives of the present paper, the authors in [20] proposed an incrementally assisted kinesthetic teaching method that refines the demonstrated trajectory in combination with DMP and dynamic time warping. The reference learned trajectory is provided to the impedance controller and the target stiffness is increased in each iteration using a predefined function. This approach increases the robot's leading role monotonically during the demonstration, without allowing the operator regain the leader's role. A smooth switching mechanism for task transition was proposed in [21] to refine the learned tasks according to the external force, by acting on the null-space or on the end-effector. In the approach we propose, a continuous adjustment in the virtual stiffness of the impedance controller is performed instead of task transition, depending on the agreement level between the demonstrations.

With the motion refinement tube proposed in [22] three areas of different stiffness values are defined depending on the magnitude of the tracking error; a high stiffness area for small tracking errors so that accurate autonomous tracking is achieved, a larger area of zero stiffness where the operator can teach deviations from the desired trajectory, and a greater perimeter of high stiffness to limit the allowed deviation. Notice that in [22], the inner high stiffness area is always present up to a maximum force, hence the operator always feels a constant attractive force towards the reference trajectory. In contrast to [22], our approach of virtual fixtures forms a tube around the reference path and not around the tracking error, assisting the operator to easily move inside the tube. In that way, virtual fixtures are temporally independent. Moreover, the tube can be penetrated with the operator feeling no force outside it, so that a completely new path can be demonstrated and not just a refinement of the previous one.

2   The Proposed Progressive Automation Framework

We consider a scenario where the operator must program a robot for a pick and place task. The objective is to design a method which will enable a novice operator to program the robot by kinesthetic demonstration and after a few demonstrations the robot can automatically carry out the task without further human assistance. The basic concept of the proposed progressive automation framework involves a robot under variable Cartesian impedance control. The task consists of movements of the robot arm demonstrated through kinesthetic teaching, as well as actions of the gripper for opening and closing the fingers of the end-effector. The change of the gripper state and the end of a task cycle are signaled using predetermined haptic cues. Each cycle consists of discrete motion segments between cues.

In the first demonstration of the task cycle, which is considered as the initialization of the system, the robot has zero stiffness and behaves passively. After each motion segment is completed, Dynamic Movement Primitives are used for compact representation of the demonstrated trajectory in the form of a dynamic model that has the properties of stable convergence to the target and scaling to new targets. Let $\mathbf{p}^i(t)$ the task coordinates of the robot that are recorded in the ith demonstration of a complete task cycle. Within the scope of this paper only the robot's position is considered. When the first task cycle is completed and the second one begins, a reference trajectory $\mathbf{p}_d$ is being provided by the DMP and is incorporated in the impedance controller via the tracking error $\tilde{\mathbf{p}} = \mathbf{p} - \mathbf{p}_d$, where $\mathbf{p}$ is the robot's current position. In parallel with the DMP, virtual fixtures are activated that assist the operator repeat the previous path with better accuracy. Virtual fixtures discourage the user to deviate significantly from the closest point $\mathbf{p}_n$ on the reference path $\mathbf{p}_{ref}$ ($\mathrm{argmin}_{\mathbf{p}_n} ||\mathbf{p}_{ref} - \mathbf{p}||$), that was previously demonstrated. The fixtures can be penetrated by the operator to allow adjustments of the reference path. The defined position errors are shown in Fig. 1 and the proposed progressive automation framework is shown in the block diagram of Fig. 2.

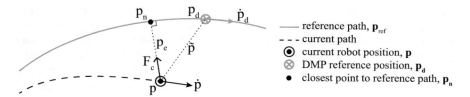

reference path, $\mathbf{p}_{\text{ref}}$
- - - current path
⊙ current robot position, $\mathbf{p}$
⊗ DMP reference position, $\mathbf{p}_d$
• closest point to reference path, $\mathbf{p}_n$

Fig. 1 Illustration of the used variables

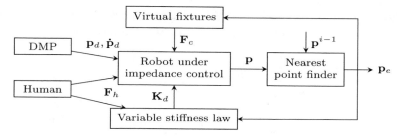

Fig. 2 Block diagram of the proposed system for progressive automation after the first demonstration ($i > 1$)

The robot under the proposed scheme is described by the following dynamic behavior in Cartesian space:

$$\mathbf{M}_d(t)\ddot{\tilde{\mathbf{p}}} + \mathbf{D}_d(t)\dot{\tilde{\mathbf{p}}} + \mathbf{K}_d(t)\tilde{\mathbf{p}} = \mathbf{F}_h - \mathbf{F}_c , \qquad (1)$$

where $\mathbf{M}_d(t)$, $\mathbf{D}_d(t)$, $\mathbf{K}_d(t) \in \mathbb{R}^{3 \times 3}$ is the target impedance, $\mathbf{F}_h \in \mathbb{R}^3$ is the inter-action force and $\mathbf{F}_c \in \mathbb{R}^3$ is the control signal implementing virtual fixtures. Notice that at zero stiffness the robot can be guided kinesthetically by the operator within the workspace, while at high stiffness the tracking error $\tilde{\mathbf{p}}$ is minimized. The variable stiffness $\mathbf{K}_d(t)$ depends on the interaction force $\mathbf{F}_h$ and the position deviation from the reference path $\mathbf{p}_e = \mathbf{p} - \mathbf{p}_n$, as described in detail in Sect. 2.1. The position deviation $\mathbf{p}_e$ is time independent as it expresses the distance of the current position of the robot from the closest point on the reference path. Matrices $\mathbf{M}_d(t)$, $\mathbf{D}_d(t)$ are calculated based on the current stiffness so that the desired dynamic behavior is critically damped. The details of each module are provided in the following sections.

2.1 Dynamic Role Allocation Through Variable Stiffness

The most dominant parameter for accurate trajectory tracking is the stiffness term $\mathbf{K}_d$. A variable stiffness strategy is proposed here that determines the appropriate stiffness value based on the external interaction force $\mathbf{F}_h$ and on the position deviation $\mathbf{p}_e$ from the reference path. When the operator demonstrates a path that is close to the reference one, the stiffness should gradually increase, allocating an increasingly leading role to the robot. The transition should be initially slow, requiring the operator to conduct a few demonstrations thus allowing fine tuning of the reference trajectory. Should a considerable divergence $\mathbf{p}_e$ be measured from the reference path, the stiffness value must remain low. On the other hand, when the robot is in a leading role (with relatively high stiffness), a high interaction force should reverse the leading role back to the operator. The rate of change in the stiffness should depend on the current stiffness value, ensuring appropriate and smooth role inversion.

The target stiffness $\mathbf{K}_d = k\mathbf{I}_3$, $k \in \mathbb{R}^+$ is derived according to the following proposed stiffness rate law that satisfies the characteristics stated above:

$$
\dot{k} = \begin{cases} max\{w, 0\}, & k = 0 \\ w, & 0 < k < k_{max}, \\ min\{w, 0\}, & k = k_{max} \end{cases} \quad with \; k(0) = 0. \tag{2}
$$

$$
w = f(k) \left(1 - \frac{||\mathbf{F}_h||}{g(k)} - \frac{||\mathbf{p}_e||}{\lambda_1} \right), \tag{3}
$$

$$
f(k) = \sqrt{k} + f_{min}, \tag{4}
$$

$$
g(k) = \lambda_2 \frac{k_c}{k + \epsilon}, 0 < \epsilon \ll 1. \tag{5}
$$

The parameter λ_1 is the threshold of $||\mathbf{p}_e||$ above which the stiffness is decreasing and is selected based on the desired accuracy. The function $g(k)$ is a variable threshold of the force norm. It depends on the current stiffness, with λ_2 being the corresponding threshold of $||\mathbf{F}_h||$ for $k = k_c$. When $||\mathbf{F}_h|| > g(k)$ then the stiffness decreases. In practice, when $k > k_c$ a high force $||\mathbf{F}_h||$ will quickly revoke the leading role to the operator. The function $f(k)$ is a scale factor, with f_{min} being the minimum rate of stiffness increase (when $k = 0$).

The rate of change w is illustrated in Fig. 3 with respect to the force and position deviation for characteristic values of k. Increase of the stiffness can only occur when both the force error and the position deviation are small. The maximum increase rate in the figure is $40 \, \frac{N/m}{s}$, while the maximum decrease rate is $-350 \, \frac{N/m}{s}$. In Fig. 3a it is shown that the decrease rate is magnified by the stiffness value (as intended with high interaction forces), thus allowing very fast leading role allocation to the operator. The effect of $||\mathbf{F}_h||$ on the stiffness rate of change and the dependency on the current stiffness is shown in the cross-section of Fig. 3b. Particularly for $k = 0$, the rate w is independent of $||\mathbf{F}_h||$. In Fig. 3c the effect of the position deviation $\mathbf{p}_e$ is shown in relation to the current stiffness and λ_1.

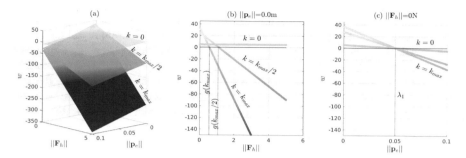

Fig. 3 Rate of variable stiffness with respect to the force and the position deviation from the reference path. 3D representation and cross-sections for $||\mathbf{p}_e|| = 0.0$ m, $||\mathbf{F}_h|| = 0$ N

2.2 Encoding the Task Cycle with DMP

The DMP framework allows the encoding of each relevant task coordinate with a compact representation [23] and consists of two dynamical systems: the transformation system and the canonical system. The transformation system is:

$$\tau^2 \ddot{y} = \alpha_y \left(\beta_y (g - y) - \tau \dot{y} \right) + f \tag{6}$$

where α_y, β_y are constants, τ is a temporal scaling factor, g is the goal, y, $\dot{y}$, $\ddot{y}$ are the position, velocity and acceleration of the DMP respectively. For each task coordinate (in this case the translational axes XYZ), the reference position $\mathbf{p}_d$ is derived as $\mathbf{p}_d = \begin{bmatrix} y_X & y_Y & y_Z \end{bmatrix}^T$. The forcing term f is given by the weighted sum of N Gaussian kernels.

$$f = \frac{\sum_{i=1}^{N} \psi_i(x) w_f^i x}{\sum_{i=1}^{N} \psi_i(x)} (g - y_0), \quad with \ \psi_i = \exp\left(-h_i (x - c_i)^2\right) \tag{7}$$

where h_i is the inverse width, c_i the center and w_f^i the weight corresponding to the ith Gaussian kernel.

The canonical system is described by an exponential decay term:

$$\tau \dot{x} = -\alpha_x x, \quad x(0) = 1 \tag{8}$$

where α_x is a time constant. To avoid high acceleration $\ddot{y}$ at the beginning of each DMP, a delayed goal attractor g is used that starts from the initial position y_0 and converges to g_d:

$$\tau \dot{g} = \alpha_g (g_d - g), \quad g(0) = y_0 \tag{9}$$

where g_d is the target of each DMP and α_g is a constant. This modification does not affect the scaling properties and stability properties of the system. Joining consecutive DMP to carry out the entire task cycle is straightforward as long as the velocity at the targets is close to zero [24]. To learn the weight parameters of the kernels, Locally Weighted Regression (LWR) is employed [25] using a single or multiple demonstrations [20]. Within the scope of this paper the first demonstration was used for determining the reference trajectory and path. Alternatively, the last demonstration or a fusion of multiple demonstrations with dynamic time warping can also be used, which would allow the user to refine the first demonstration.

2.3   Virtual Fixtures for User Assistance

In order to assist the user along the reference path, virtual fixtures are utilized based on a virtual potential of the scalar distance between the end-effector and the nearest point $\mathbf{p}_n$ to the reference path. By establishing such a potential, a guidance tube is formed around that path. In that way, the assistance provided to the operator by the virtual fixtures is independent from the value of the virtual stiffness of the impedance controller, that is responsible for tracking the reference trajectory and hence, it is particularly useful at the initial demonstrations when $\mathbf{K}_d$ is small.

Let us define the scalar value $d \in \mathbb{R}^+$ as a representative metric of the minimum distance from the path as:

$$d = \frac{||\mathbf{p_e}||^2}{r^2} \tag{10}$$

with $r \in \mathbb{R}^+$ being the radius of the virtual fixtures tube. Our proposed virtual fixture control signal $\mathbf{F}_c$ is motivated by the potentials utilized in [26], acts only within the tube ($d < 1$) and vanishes outside it ($d > 1$). More specifically, when approaching the walls of the tube, smooth corrective repulsive forces should be applied to the user towards the center. The walls of the tube should be penetrated when the forces exceed a desired threshold. Outside the tube no constraining force should be applied but reinstatement within the tube should be allowed.

To achieve our control objective, we introduce the following transformation:

$$T(d) = \begin{cases} 10d^3 - 15d^4 + 6d^5 , 0 \le d \le 1, \\ \qquad\qquad 1 \qquad\qquad , d > 1. \end{cases} \tag{11}$$

and we define the following control signal:

$$\mathbf{F}_c = -a_c \nabla U(d) = -a_c \frac{\partial T}{\partial d} T(d) \mathbf{p_e}. \tag{12}$$

where $U = \frac{1}{2}T^2(d) \in \mathbb{R}^+$ represents the virtual potential and a_c is a positive control gain. Notice that the potential's global minimum is on the reference path. Thus, the

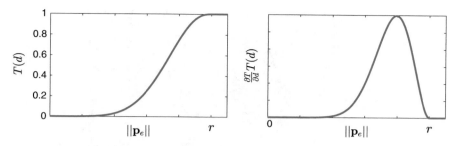

Fig. 4 Potential field and its gradient for virtual fixtures

constraint force $\mathbf{F}_c$ is zero when $\mathbf{p}$ lies on the path (since $\mathbf{p}_e = \mathbf{0}$) or when it lies outside the tube at $||\mathbf{p}_e|| > r$ (since $\frac{\partial T}{\partial d} = 0, \forall d > 1$). Figure 4 depicts the potential and the gradient as a function of $||\mathbf{p}_e||$.

3 Experimental Evaluation

To evaluate the proposed framework in a realistic scenario, the PbD of the pick and place task shown in Fig. 5 is selected. The objective is to kinesthetically guide the robot to pick objects from the bottom of the automated part feeder and place them over the container, until the robot is able to carry out the task autonomously.[1] A KUKA LWR 4+ manipulator is used with a BH8 Barrett hand as the end-effector.

The impedance controller of Eq. (1) is implemented without inertia shaping and with a locked orientation of the end-effector. $\mathbf{D}_d$ is determined according to the current $\mathbf{K}_d$ and the robot's inertia matrix. The cues are provided by a simple classifier using the external torque measurement of the fingers, while the robot's external force estimation is used to obtain $\mathbf{F}_h$. The DMP parameters for Eqs. (6)–(9) are $\alpha_y = 20$, $\beta_y = \alpha_y/4$, $\alpha_x = -ln(0.001)$, $\alpha_g = 5$, and 50 Gaussian kernels are utilized for each segment. Training of the DMP is performed in the first demonstration of the task cycle. The variable stiffness is calculated by Eqs. (2)–(5) using $k_{max} = 2000$ N/m, $\lambda_1 = 0.05$, $\lambda_2 = 6$, $k_c = 200$ N/m, $f_{min} = 5 \frac{\text{N/m}}{\text{s}}$ and the virtual fixtures of Eqs. (10)–(12) using $r = 0.01$ m and $\alpha_c = 1500$. The operation of the gripper switches to autonomous mode (opening/closing) whenever $k = k_{max}$. A DMP finishes and the next one begins when it reaches the goal position within an accuracy of 10^{-4} m.

An operator is asked to program the robot by demonstrating the task with the proposed progressive automation framework. The operator is informed of the task objective and that the robot will be able to perform the task autonomously after an unknown number of demonstrations. When the operator is confident that the robot can perform the task without further assistance, he is encouraged to stop interacting with it. For the selected pick and place task two cues per task cycle are required,

---

[1] Video demonstration: https://youtu.be/1m185Z9y5JI.

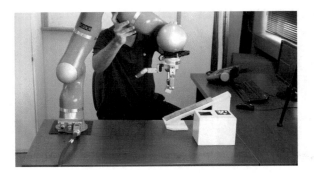

Fig. 5 Use-case of progressive automation in a pick and place task

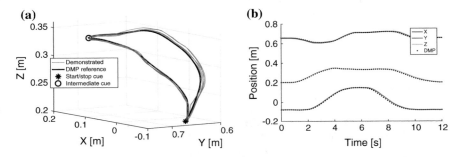

Fig. 6 **a** Robot trajectory learned during progressive automation and **b** the DMP approximation of the first demonstrated task cycle

that segment the demonstration into an equal number of parts, as it is shown in Fig. 6a. This figure illustrates the task coordinates of the robot's end-effector during the progressive automation in each task cycle. The reference trajectory produced by the DMP accurately approximates the first demonstration of the task cycle with a root mean square error of the position less than 0.005 m as shown in Fig. 6b.

The proposed variable stiffness law adjusts the target stiffness of the impedance controller as shown in Fig. 7a. To demonstrate the effectiveness of the variable stiffness law in combination with the proposed virtual fixtures, a comparison of the same task is presented where the operator repeats the procedure with virtual fixtures being turned off so that they do not provide assistance. It can be inferred from the figure that when virtual fixtures are enabled, the stiffness law increases k to it's maximum value after four demonstrations. On the contrary, eight demonstration are required when virtual fixtures are disabled, mainly due to the inherent human variability which prevents accurate repetition of the previous demonstration. In both cases, however, the stiffness k starts to increase when the first task cycle is completed with an increasingly higher rate according to the properties of the proposed variable stiffness function described at Sect. 2.1.

The interaction force norm $||\mathbf{F}_h||$ presented in Fig. 7b shows that when virtual fixtures are enabled, the operator stops interacting with the robot at the end of the

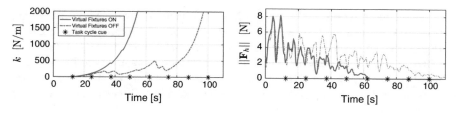

Fig. 7 Stiffness and force variation with and without virtual fixtures. Asterisks represent the time instants when a task cycle is completed (using a haptic cues). Progressive automation with virtual fixtures is achieved after 4 demonstrations

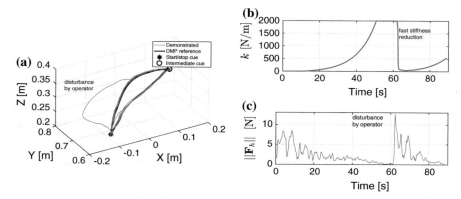

Fig. 8 Fast stiffness reduction when a high external force is applied by the operator (at t = 62 s)

fifth demonstration (t = 60 s). The provided effort of the operator presents a reduction trend at each task cycle, which is expected as the variable stiffness approach gradually increases the leading role of the robot. Without virtual fixtures, significantly more demonstrations are required.

After the operator has demonstrated the task and the robot is in autonomous mode, the operator induces a disturbance in order to illustrate the case of intentional adjustment to the demonstrated trajectory as shown in Fig. 8a. The applied force at t = 62 s (Fig. 8c) that is approximately 12 N at peak, yields a rapid reduction of the stiffness, safely reverting the leading role to the operator (Fig. 8b). In such cases, the latest demonstration can be utilized to retrain the DMP.

4 Conclusions

The main target of this paper is to introduce a framework that allows an operator to intuitively program a robot for repetitive tasks through kinesthetic teaching and seamless transition to autonomous operation. The proposed progressive automation framework aims at utilizing the capabilities of collaborative robots for efficient and fast programming of tasks in an industrial or a domestic environment. The validity of the proposed approach is given in a pick and place task from a single operator. Apart from point to point movements, the method is applicable to trajectory following tasks. Overall, the programming of the investigated task can be completed with progressive automation in less than a minute, without any interaction with a user interface or pre-programming of the task cycle sequence.

Future work includes further development of the framework by including the end-effector's orientation, by enhancing the DMP implementation, and by evaluating its user friendliness via an extensive user study with qualitative and quantitative criteria.

Acknowledgements Fotios Dimeas is funded by "IKY Fellowships of Excellence for Postgraduate Studies in Greece—Siemens Program".

References

1. Pan, Z., Polden, J., Larkin, N., Van Duin, S., Norrish, J.: Recent progress on programming methods for industrial robots. Robot. Comput.-Integr. Manuf. **28**(2), 87–94 (2012)
2. Schou, C., Damgaard, J.S., Bogh, S., Madsen, O.: Human-robot interface for instructing industrial tasks using kinesthetic teaching. In: 4th International Symposium on Robotics, ISR 2013 (2013)
3. Billard, A.G., Calinon, S., Dillmann, R.: Learning from humans. In: Springer Handbook of Robotics. Springer International Publishing, pp. 1995–2014 (2016)
4. Akgun, B., Cakmak, M., Yoo, J., Thomaz, A.: Trajectories and keyframes for kinesthetic teaching: a human-robot interaction perspective. In: International Conference on Human-Robot Interaction, pp. 391–398 (2012)
5. Schroecker, Y., Amor, H.B., Thomaz, A.: Directing policy search with interactively taught via-points. In: International Conference on Autonomous Agents and Multiagent Systems, pp. 1052–1059 (2016)
6. Steinmetz, F., Montebelli, A., Kyrki, V.: Simultaneous kinesthetic teaching of positional and force requirements for sequential in-contact tasks. In: IEEE-RAS International Conference on Humanoid Robots, pp. 202–209, Dec 2015
7. Skoglund, A., et al.: Programming by demonstration of pick-and-place tasks for industrial manipulators using task primitives. In: International Symposium on Computational Intelligence in Robotics and Automation, pp. 368–373. IEEE (2007)
8. Lambrecht, J., Kleinsorge, M., Rosenstrauch, M., Krüger, J.: Spatial programming for industrial robots through task demonstration. Int. J. Adv. Robot. Syst. **10**(5), 254 (2013)
9. de Rengerve, A., Hirel, J., Andry, P., Quoy, M., Gaussier, P.: On-line learning and planning in a pick-and-place task demonstrated through body manipulation. In: 2011 IEEE International Conference on Development and Learning, pp. 1–6 (2011)
10. Dean-Leon, E., Ramirez-Amaro, K., Bergner, F., et al.: Robotic technologies for fast deployment of industrial robot systems. In: 42nd Annual Conference of the IEEE Industrial Electronics Society, pp. 6900–6907. IEEE (2016)
11. Groth, C., Henrich, D.: One-shot robot programming by demonstration using an online oriented particles simulation. In: 2014 IEEE International Conference on Robotics and Biomimetics, IEEE ROBIO 2014, pp. 154–160 (2014)
12. Evrard, P., Kheddar, A.: Homotopy-based controller for physical human-robot interaction. In: RO-MAN 2009—The 18th IEEE International Symposium on Robot and Human Interactive Communication, pp. 1–6 (2009)

13. Jarrasse, N., Sanguineti, V., Burdet, E.: Slaves no longer: review on role assignment for human-robot joint motor action. Adapt. Behav. (2013)
14. Kucukyilmaz, A., Sezgin, T., Basdogan, C.: Intention recognition for dynamic role exchange in haptic collaboration. IEEE Trans. Haptics 6(1), 58–68 (2013)
15. Medina, J.R., Lawitzky, M., Mörtl, A., Lee, D., Hirche, S.: An experience-driven robotic assistant acquiring human knowledge to improve haptic cooperation. In: IEEE International Conference on Intelligent Robots and Systems, pp. 2416–2422 (2011)
16. Li, Y., Tee, K.P., Chan, W.L., Yan, R., Chua, Y., Limbu, D.K.: Continuous role adaptation for human-robot shared control. IEEE Trans. Robot. 31(3), 672–681 (2015)
17. Ficuciello, F., Villani, L., Siciliano, B.: Variable impedance control of redundant manipulators for intuitive human-robot physical interaction. IEEE Trans. Robot. 31(4), 850–863 (2015)
18. Dimeas, F., Aspragathos, N.: Online stability in human-robot cooperation with admittance control. IEEE Trans. Haptics 9(2), 267–278 (2016)
19. Bussy, A., Gergondet, P., Kheddar, A., et al.: Proactive behavior of a humanoid robot in a haptic transportation task with a human partner. In: IEEE International Workshop on Robot and Human Interactive Communication, pp. 962–967 (2012)
20. Tykal, M., Montebelli, A., Kyrki, V.: Incrementally assisted kinesthetic teaching for programming by demonstration. In: ACM/IEEE International Conference on Human-Robot Interaction, pp. 205–212, Apr 2016
21. Saveriano, M., An, S.I., Lee, D.: Incremental kinesthetic teaching of end-effector and null-space motion primitives. In: Proceedings—IEEE International Conference on Robotics and Automation, pp. 3570–3575, June 2015
22. Lee, D., Ott, C.: Incremental kinesthetic teaching of motion primitives using the motion refinement tube. Auton. Robots 31, 115–131 (2011)
23. Ijspeert, A.J., Nakanishi, J., Hoffmann, H., Pastor, P., Schaal, S.: Dynamical movement primitives: learning attractor models for motor behaviors. Neural Comput. 25(2), 328–373 (2013)
24. Kulvicius, T., Ning, K., Tamosiunaite, M., Wörgötter, F.: Joining movement sequences: modified dynamic movement primitives for robotics applications exemplified on handwriting. IEEE Trans. Robot. 28(1), 145–157 (2012)
25. Schaal, S., Atkeson, C.: Constructive incremental learning from only local information. Neural Comput. 10(8), 2047–2084 (1998)
26. Karayiannidis, Y., Droukas, L., Papageorgiou, D., Doulgeri, Z.: Robot control for task performance and enhanced safety under impact. Front. Robot. AI 2, 1–12 (2015)

Part IV
Robot Learning

Enhancing Shared Control via Contact Force Classification in Human-Robot Cooperative Task Execution

Jonathan Cacace, Alberto Finzi and Vincenzo Lippiello

Abstract In this paper, we present a novel method to support physical human-robot interaction during the execution of collaborative manipulation tasks. In the proposed approach, the robot is able to infer the operator intentions from the human contact forces, exploiting such information to properly react to the operator interventions and suitably adapt the execution of the shared task. In particular, we assume that the robotic system can autonomously generate and execute Cartesian trajectories, while a human operator can provide interventions exerting contact forces on the robot itself. The resulting robot motion is obtained by mixing in an adaptive manner the input commands provided by both the robotic control system and the human operator. In our approach, human intention estimation relies on a Neural Network capable of distinguishing the operator contact forces that support or oppose the autonomous motion planned by the robotic system. We tested the system at work in different scenarios considering simple interaction tasks performed with the 7-DOF Kuka LWR IIWA manipulator and comparing the performance obtained by a human operator with and without the assistance of the proposed system. The collected results demonstrate the effectiveness of the proposed approach.

Keywords Human-robot cooperation · Collaborative robotics · Intuitive
interfaces · Shared control

J. Cacace · A. Finzi (✉) · V. Lippiello
Department of Electrical Engineering and Information Technology,
Università degli Studi di Napoli Federico II, Via Claudio 21, 80125 Naples, Italy
e-mail: alberto.finzi@unina.it

J. Cacace
e-mail: jonathan.cacace@unina.it

V. Lippiello
e-mail: vincenzo.lippiello@unina.it

© Springer International Publishing AG, part of Springer Nature 2019   167
F. Ficuciello et al. (eds.), *Human Friendly Robotics*, Springer Proceedings
in Advanced Robotics 7, https://doi.org/10.1007/978-3-319-89327-3_13

1   Introduction

The capability of properly reacting to human physical interventions is a critical issue in human-robot collaborative manipulation. Physical interaction is indeed a typical an natural cooperation mode when humans and robots are involved in the execution of shared tasks [4], as exemplified by the case of an operator that co-works with lightweight arms in industrial scenarios. Apart from safety aspects, one fundamental problem for an effective collaboration between humans and robots is to take into account human intentions during the interaction process. Indeed, a system that is capable of inferring, predicting, and exploiting the human intentions during the interaction can enhance the effectiveness of the human-robot cooperation both from the human and the robotic side [8]. In this direction, we present a shared control framework that permits to infer the operator's intentions during physical human-robot co-manipulation in order to suitably regulate and adapt the robot execution and interaction mode. In the proposed system, we assume that the robotic system is able to autonomously generate and execute Cartesian trajectories for the execution of a given task, while the human operator can physically interact with a robotic manipulator in order to on-line adjust, speed up, or slow down the planned motions. In this scenario, we propose an approach where the operator physical interventions during the interaction are continuously classified over different intentions, in order to assess whether the human guidance is supporting or opposing the planned robot motion. The estimated intentions are then exploited to regulate the robot attitude during task execution, which can range from an autonomous and proactive mode to a passive one, when the human interventions seems not aligned with the planned targets. The operator intentions are recognized exploiting a *3-Layer Neural Networks* that assesses input data about the operator contact forces and the actual robot motion, inferring the the operator intention to: follow the planned path, deviate from the latter, or retrace part of the path already covered. We tested the effectiveness of the proposed system with experiments on a 7-DOF Kuka *LWR IIWA* manipulator endowed with an ATI *Mini45* force sensor connected to the last joint of the robot (see Fig. 1). The overall system has been tested in a simple collaborative task, in which different users are to physically interact with the robotic system in order to move the end effector along a desired path, while performing some prescribed deviations. In this context, we assessed the performance of our framework with respect to alternative simpler approaches. The collected results show the advantage the proposed solution in terms of both task performance and human effort reduction.

The rest of the paper is organized as follow. In Sect. 2 a brief overview of the state of art on intention estimation in human-robot cooperation is presented. Section 3 describes the enhanced shared control architecture proposed in this work. Intention classification is illustrated in Sect. 4, while Sect. 5 describes how the operator's intentions are exploited for collaborative task execution. In Sect. 6 different test cases are introduced and discussed, while conclusions are provided in Sect. 7.

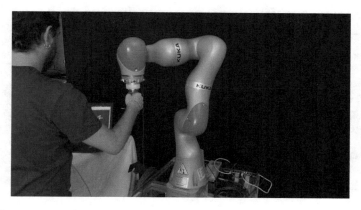

Fig. 1 Human operator interacting with a Kuka LBR IIWA robot endowed with an ATI mini-45 force sensor

2 Related Works

In human-robot collaboration, intention estimation and anticipatory processes are considered as crucial enabling mechanisms for the interaction fluency and effectiveness [8, 12]. As for physical human-robot interaction, different approaches have been proposed in order to enable a human aware and compliant interaction. For instance, in [21], the authors propose a method to on-line regulate the robot physical behavior with respect to the estimated human fatigue. More related to our work, in [10] a shared trajectory generator, based on the operator force contacts, is proposed to translate the human intentions into ideal trajectories for the robot. Here, an on-line trajectory generator is integrated with a classical impedance control system, but an intention estimation system is not deployed. Similarly, [16] proposes a framework for switchings between trajectory tracking and force minimization in physical human-robot collaboration, but also in this case, intention recognition is not considered. In [5], the robotic control schema is regulated according to the human physical interaction, which is observed as a change in control effort. Intention recognition methods typically consider external forces excreted by the human operator on the robot side [5, 7, 13, 18–20]. Similarly to our approach, authors in [15] exploit the operator force contacts to adapt the robot behavior during the execution of shared tasks. However, differently from our approach, in this case a game theoretic solution is proposed. In [14], the motion intention of the human partner is detected using the human limb model to estimate the desired trajectory; instead, in [11], external forces are exploited in order to discern between an intentional human contact and an unexpected collision. Human motion estimation is also deployed in [6], where the authors exploit Neural Networks to extract human motion parameters and then predict whether the human interventions are active or passive. In contrast, we exploit Neural Networks to directly classify the human force contacts with respect to the robot motion during the execution of a cooperative task.

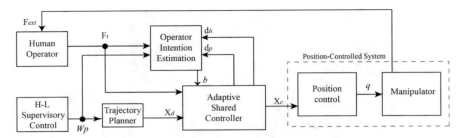

Fig. 2 Adaptive shared control architecture

3 Control Architecture

In this section, the proposed control architecture is detailed (see the block schema in Fig. 2). The core of the proposed system is the *Operator Intention Estimation* module, whose aim is to infer the operator intentions taking into account the autonomous motion of the robot along with the motion induced by the operator interventions. The output of this module is provided to the *Adaptive Shared Controller* module, which is responsible of the overall robot control since it combines the human and the robotic contributions to the motion. In the following, we further detail the architecture describing the associated modules.

The human operator physically interacts with the robot using a force torque sensor mounted on the manipulator end effector. This way, the operator is able to exert a force F_t on the manipulator moving it within its workspace. On the other end, the operator is influenced by the force feedback F_{ext} generated by the low-level controller of the robot. The *H-L Supervisory Control* module guides the robot end effector through a waypoint navigation by incrementally selecting the next position (Wp) the manipulator is to reach. The selected waypoint is then sent to a *Trajectory Planner* module, which generates a Cartesian trajectory to move the end effector towards the selected destination, streaming the desired motion X_d of the robot. As for the low-level control of the robot, we assume to directly command the end effector position, relying on the robot inner control loop to solve the associated inverse kinematic problem (*Position-Controlled System*). The desired position is generated by the *Adaptive Shared Controller* module, which is responsible for the combination of the motion X_d data, autonomously planned by the robot, with the control inputs F_t provided by the operator, in order to generate a *compliant* position command X_c for the *Position-Controlled System*. This is obtained through an admittance controller that establishes a dynamic relationship between the forces applied to the robot and the displacement from its desired position [9]. The admittance controller is described by the typical second-order relationship:

$$m\ddot{x} + d\dot{x} + kx = F \tag{1}$$

that in our case can be expressed as:

$$M_d(\ddot{X}_c - \ddot{X}_d) + D_d(\dot{X}_c - \dot{X}_d) + K_d(X_c - X_d) = F_t \tag{2}$$

with M_d, D_d and K_d representing the desired virtual inertia, virtual damping and virtual stiffness, respectively. The admittance controller enables the operator to provide interventions that slightly modify and adjust the planned trajectory during its execution. In this setting, the estimation of human intentions is exploited to suitably adapt the robot behaviour according to the operator force contact. This way, depending on the estimated intention of the operator, the robotic system can switch from a proactive behaviour, that guides the human, to a passive one, in which the human leads the interaction. This mechanism is better described in Sect. 5. Human intention estimation is performed by the *Operator Intention Estimation* module exploiting a *3-Layer Neural Network* to assess whether the operator intentions are aligned/misaligned with respect to the planned robotic path/motion. Specifically, intention estimation relies on input data provided by both the human and the robot guidance, such as: the desired motion direction d_d, the current motion direction requested by the operator d_h, the target point to reach Wp, and the forces exerted on the robot end effector F_{ext}. The classification process along with the associated adaptive behaviour of the robot will be better detailed in the next sections.

4 Human Intention Estimation

This section describes the operator intention recognition process. As already stated, the recognition process relies on a *Neural Network* classifier [2]. In particular, we deployed a multi-layer feed-forward neural network. The proposed Neural Network (NN) is introduced to classify the human operator intentions with respect to the motion autonomously planned and executed by the robot. In this respect, the aim is to assess weather the intention behind the operator physical guidance is aligned or not with respect to the current robotic target and motion. As reported in Table 1, we distinguish the human intentions in four classes: the human intention is aligned with the current robot behavior (*Coincident*); the human is adjusting the trajectory, but the target remains aligned with the estimated operator intent (*Coincident diverging*); the human is adjusting the trajectory with a movement that is not coherent with the target (*Opposite diverging*); the human intention is misaligned with respect to the robot motion and target (*Opposite*). The proposed NN is three-layered. The output layer contains 4 nodes, one for each class of recognized intentions. As for the middle (hidden) layer, we introduce 25 nodes. Finally, we have 3 input nodes associated with the following data:

- $||F_t||$: The magnitude of forces exerted by the operator.
- $||X_c - C_p||$: The distance between the current end effector position and the closest point on the planned trajectory.
- $\angle(\vec{d_c}, \vec{d_d})$: The deviation between the planned motion and the human motion, represented by the angle between the movement vectors.

Table 1 Operator intentions classes

Class ID	Class name	Description
#0	*Coinciding*	The operator moves the manipulator following the planned path
#1	*Opposite*	The operator moves the manipulator in the direction opposite to the planned path
#2	*Coinciding deviation*	The operator deviates from the planned path, trying to reach the target
#3	*Opposite deviation*	The operator deviates from the planned path, moving towards a direction opposite and distant from the target

Here, the force exerted by the operator F_t is used to calculate the human motion direction d_c, while the end effector Cartesian velocity is the planned motion direction d_d. Finally, C_p represents the closest point between the end effector and the planned trajectory. Depending on these values, we classify the human intention. An exemplification of our approach is illustrated in Fig. 3, where we provide an intuitive representation of the input values in the 4 cases. Figure 3a shows the *Coinciding* situation, where the operator moves the end effector along the planned path. Similarly, Fig. 3c provides an example of a *Coinciding deviation*: the end effector is deviated from the path, but the motion remains coherent with respect to the target. In contrast, Fig. 3b and 3d show the cases in which, respectively, the human interventions are *Opposite* to the target or deviating from the planned path, and not oriented towards the target (*Opposite deviation*).

Training. In order to train the NN, we involved a group of ten users (students and researchers) not aware about the system functioning, asking them to interact with a robotic system while it was generating and executing random linear trajectories. For each user, the training session lasted approximately 10 min. In this experimental setting, we collected several examples of the operator interactive behaviour for each class of intentions. The resulting training set, which contains about 12000 samples, was then exploited for NN training deploying a *back propagation* algorithm.

Testing. We tested the system with a different group of ten users (students and researchers). Each tester was requested to interact with the manipulator during the execution of a cyclic squared path providing a set of predefined interventions. In particular, each user performed four different evaluation sessions, adopting a different predefined behavior during the interaction with the robot. Table 2 reports the *confusion matrix* of the intention classifier, in which, each row of the matrix represents the instances of a predicted class, while each column represents the instances of an actual class. The collected results show that the proposed recognition system achieves satisfactory results for our collaborative control purpose.

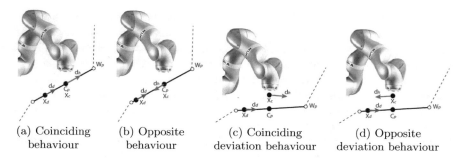

(a) Coinciding behaviour (b) Opposite behaviour (c) Coinciding deviation behaviour (d) Opposite deviation behaviour

Fig. 3 Example of different operator intention classification

Table 2 Confusion matrix

	0 (%)	1 (%)	2 (%)	3 (%)
0	91.6	0.23	8.17	0
1	0.28	70.02	6.9	22.8
2	11.2	0.32	82.4	6.08
3	11.34	0.99	16.02	71.63

5 Adaptive Shared Control

In this section, we discuss how the *Adaptive Shared Controller* of the robot exploits the human operator intentions in order to regulate the execution of the collaborative manipulation. In this setting, depending on the interpretation of the human physical interventions, the robotic system can switch from a *passive* to an *active* operative mode. In the first case, the manipulator is fully compliant to operator force contact, without any robotic contribution towards the desired motion. Instead, in the second case, the robotic system can generate and execute motion trajectories to proactively assist the operator in reaching the target positions. Considering the the control architecture depicted in Fig. 2, the target position is a state: $Wp = (x, y, z)$; in order to reach this position, the *Trajectory Planner* generates a cartesian trajectory in terms of position, velocity, acceleration, and jerk. The trajectory is generated exploiting a 4th order spline concatenation method that preserves continuous acceleration. In particular, at each time step, a reference command $X_d = (x_d, \dot{x}_d, \ddot{x}_d)$ is generated to actuate the robot according to the admittance control formula (see Eq. 2). This way, the compliant position of the end effector is obtained by integrating both the operator and the robot control data:

$$\ddot{X}_{c_{i+1}} = \frac{M\ddot{X}_{d_i} + D(\dot{X}_{d_i} - \dot{X}_{c_i}) + K(X_{d_i} - X_{c_i}) + F_t}{M} \tag{3}$$

We can distinguish four different behaviours implemented by the *Adaptive Shared Controller*, one for each class recognized by the *Operator Intention Estimation* module. Specifically, in the case of a *Coinciding* intention (the operator intention is aligned with the robot plan), the manipulator will be attracted towards the next target point in the path. Differently, in the case of an *Opposite* intention, the robot is to switch to a passive mode in order to follow the human guidance. In this mode, the virtual stiffness from the previous equation is removed and the desired acceleration and velocity are set to zero. More specifically, by setting $\ddot{X}_{d_i} = \dot{X}_{d_i} = 0$ and $K = 0$, we inhibit the motion contribution due to the execution of the planned path. During this mode, the robotic system can also try to infer the new human target and replan accordingly; a mechanism of this kind is proposed in a companion paper. In the case of *Coinciding deviation*, the robotic manipulator should smoothly guide the human towards the planned trajectory, while attracting him/her towards the destination point. For this purpose, the adaptive shared controller selects as the next waypoint the midpoint Mp between the actual closest point C_p (on the planned trajectory) and the target point W_p (see Fig. 4a). This way, the operator guidance is combined with a robotic guidance towards the planned trajectory, closer to the destination point. In contrast, when the estimated behaviour is *Opposite deviation*, the operator seems to deviate from the planned path and not approaching the current target. On the other hand, the human intention is not classified as *Opposite*, hence the current target and trajectory can be maintained as a possible guidance until another intention is estimated (e.g. *Opposite* or *Coinciding deviation*). In this mode, the *Adaptive Shared Control* attracts the end effector towards the planned trajectory by providing the closed point Cp as the next waypoint to reach (see Fig. 4b). If the end effector reaches the planned trajectory during an *Opposite deviation* mode, then the next waypoint directly becomes the target point. Notice that, when the operator deviates from the planned trajectory (cases 2 and 3 of Table 1), the proposed approach provides the human with a force feedback and a consequent physical perception of the deviation between the current position of the manipulator and the planned path. The effectiveness of this feedback during the human guidance has been already proposed and tested by the authors in the context of haptic teleoperation of aerial vehicles [1, 3]; in this work a similar concept is deployed for cooperative manipulation tasks. In order to avoid control input discontinuities that may induce instability, we introduce a smooth transition between the control modes. Specifically, following the approach by [17], when a variation of the operator intent is detected, we filter the velocity command through a time-vanishing *smoothing term*. For example, in the case of a mode switch at time $t = 0$, the velocity command is computed as follows:

$$v(t) = v_a(t) + e^{1/\gamma}(v_p(0) - v_a(0)) \tag{4}$$

where γ is a constant that regulates the duration of the transition phase, while v_a and v_p are the velocity commands before and after the switch, respectively.

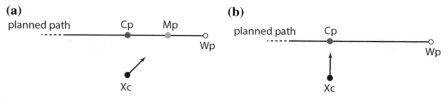

Fig. 4 Attraction points during coinciding deviation (*left*) and opposite deviation (*right*)

6 Case Study

In order to demonstrate the effectiveness of the proposed system, two different test cases have been defined distinguishing between a *planar* and a *non-planar* scenario. These test cases are further detailed below. For each scenario, we asked to a group of users to perform a manipulation task, where the user is to drive the manipulator along a desired path, displayed on a dedicated user interface, and then diverge from this path in order to reach some virtual landmarks that appear on the interface during task execution. We compared the users performance considering three different interaction modalities:

- *Manual*: during the interaction the robot is not provided with a path to execute, hence the user moves the manipulator without any guidance from the robot, which is passive and compliant to the human contact;
- *Shared*: the robot can autonomously execute the desired path, while the human can physically interact in order to deviate the trajectory. The human and the robot contributions are here directly integrated through an admittance controller, but the operator intention estimation is disabled;
- *Fully assisted*: the shared controller exploits the intention estimator to adapt the execution as proposed in this work.

We asked a group of 10 users (students and researchers) to perform 5 tries of both tests in the 3 interaction modalities described above. For each try, the user is to interact with the manipulator in order to complete the proposed path as fast as possible. During the execution, the manipulator end effector (assisted by the human) is to navigate through all the waypoints, reaching most of the virtual landmarks, while minimizing the distance form the planned path. In order to evaluate the user performance, the following variables are monitored:

- *Rewards*: Number of reached landmarks.
- *Time*: Completion time of the test.
- *Trajectory Adherence*: Distance from the planned path.
- *Path Length*: Total distance covered by the robot end effector.
- *Contact Force*: The force exerted by the operator to interact with the robot.

Each user performed the test cases in different orders, counterbalancing the experiments in different interaction modality to address the problem of learning effect. In

(a)   **(b)**

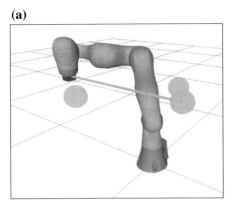   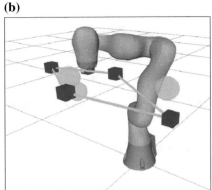

Fig. 5 User interface showing the desired path and the associated landmarks, respectively, in test 1 (**a**) and test 2 (**b**)

order to make the users aware about the task progress, the position of the robot end effector, the planned path an the location of landmarks are also shown on a dedicated user interface (see Fig. 5). In the following, we present the results collected in the two test settings.

Planar Scenario. In this scenario, the operator is to move the robot end effector along a desired path, which consists of one waypoint only. In this test case, the path is *planar*, hence the operator should not change the height of the end effector pose, which should be moved along a linear path of about 70 cm (see Fig. 5a). Three virtual landmarks appears at the start of each test, displaced away from the path, but embedded in the same plane. In this setting, the operator is to interact with the manipulator (deviating from the predefined trajectory) in order to reach a maximum number of landmarks. A landmark is considered as reached when the end effector distance is below a fixed threshold. The final reward is the number of reached landmarks.

Non-Planar Scenario. In the second scenario (*non-planar scenario*) the end effector should move along a circular path. In this setting, the altitude of the robot end effector can change (non-planar case). In addition, in contrast with the planar test case, not all landmarks appear at the start of the experiment, but some of them pop up in the user interface when the end effector passes through predefined positions during task execution. The path to cover consists of four waypoints and the total path length is 1.45 m. Analogously to the previous test case, the reward is the number of reached landmarks.

Result Analysis. Tables 3 and Table 4 provide the mean values of the data (reward, time, path length, trajectory adherence, and contact forces) collected in the three operative modes for the *planar* scenario and the *non-planar* one, respectively. In both these test scenarios, the *Assistance* mode achieves higher performance than the others in terms of gained reward, path length, trajectory adherence, and forces exerted by operators (and consequently his/her fatigue). In particular, we observed a clear improvement in adherence, indeed, in the *Assistance* mode, the testers seem pretty

Table 3 Test 1

Operative mode	Reward	Time (s)	Path (m)	Adherence (m)	Force (N)
Manual	2.5/3	11.1	1.21	0.7	45.05
Shared	2.5/3	10.4	1.22	0.23	39.3
Assistance	3/3	9.5	1.18	0.19	21.54

Table 4 Test 2

Operative mode	Reward	Time (s)	Path (m)	Adherence (m)	Force (N)
Manual	4/5	23.7	2.3	0.36	47.05
Shared	4.0/5	31.6	2.19	0.23	41.1
Assistance	4.5/5	26.9	1.8	0.13	28.7

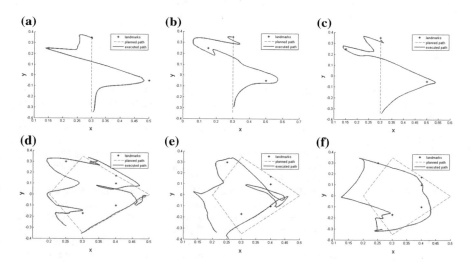

Fig. 6 Trajectories of the end effector (*blue line*) in the planar (*a, b, c*) and non-planar scenario (*d, e, f*) in *manual*, *shared* and *assisted* mode. The desired paths are the *red dashed lines*, while the landmarks are the *black stars*

effective in tracking the desired path during cooperative manipulation. This is exemplified by Fig. 6 that illustrates examples of performed tests in the 3 modes (manual, shared, assisted) for the *planar* (Fig. 6a–c) and *non-planar* scenarios (Fig. 6d–f). For each case, the desired planned paths is reported (red dashed line) along with the test landmarks (black stars) and the performed path (the blue line). As for time to accomplish the task, we could not observe an improvement in the *Assisted* mode, but this is mainly due to the slow robot velocity during autonomous execution.

7 Conclusions

We presented an approach to physical human-robot interaction during cooperative manipulation tasks, based on intention estimation. In the proposed framework, the robotic system is able to classify the operator intentions from the human force exerted on the manipulator end effector during task execution. The estimated intentions are then exploited by an adaptive shared control system to regulate its guidance towards a planned path. We described the overall framework and discussed its performance in a case study designed to compare the proposed system with respect to simpler alternatives. The collected results show the advantage of the approach in smoothly mixing the human and the robot guidance during a collaborative manipulation, as illustrated by the higher task performance and the reduced human effort with respect to the considered alternatives. These promising experimental results encourages us to deploy and test the proposed framework in more complex human-robot cooperative tasks, such as a cooperative assembly or a high weight transportation tasks.

References

1. Bevacqua, G., Cacace, J., Finzi, A., Lippiello, V.: Mixed-initiative planning and execution for multiple drones in search and rescue missions. In: Proceedings of ICAPS 2015, pp. 315–323. AAAI Press (2015)
2. Bishop, C.M.: Neural Networks for Pattern Recognition. Oxford University Press Inc., New York, NY, USA (1995)
3. Cacace, J., Finzi, A., Lippiello, V.: A mixed-initiative control system for an aerial service vehicle supported by force feedback. In: 2014 IEEE/RSJ International Conference on Intelligent Robots and Systems, pp. 1230–1235 (2014)
4. Corrales, J.A., Garcia Gomez, G.J., Torres, F., Perdereau, V.: Cooperative tasks between humans and robots in industrial environments. Int. J. Adv. Robot. Syst. 9(3), 94–104 (2012)
5. Erden, M.S., Tomiyama, T.: Human-intent detection and physically interactive control of a robot without force sensors. Trans. Robot. 26(2), 370–382 (2010)
6. Ge, S.S., Li, Y., He, H.: Neural-network-based human intention estimation for physical human-robot interaction. In: 2011 8th International Conference on Ubiquitous Robots and Ambient Intelligence (URAI), pp. 390–395 (2011)
7. Gribovskaya, E., Kheddar, A., Billard, A.: Motion learning and adaptive impedance for robot control during physical interaction with humans. In: 2011 IEEE International Conference on Robotics and Automation, pp. 4326–4332 (2011)
8. Hoffman, G., Breazeal, C.: Effects of anticipatory action on human-robot teamwork efficiency, fluency, and perception of team. In: Proceeding of the ACM/IEEE International Conference on Human-Robot Interaction—HRI '07, pp. 1–8 (2007)
9. Hogan, N.: Impedance control: an approach to manipulation. In: IEEE American Control Conference, pp. 304–313 (1984)
10. Jlassi, S., Tliba, S., Chitour, Y.: An online trajectory generator-based impedance control for co-manipulation tasks. In: IEEE Haptics Symposium, HAPTICS, pp. 391–396 (2014)
11. Kouris, A., Dimeas, F., Aspragathos, N.: Contact distinction in human-robot cooperation with admittance control. In: 2016 IEEE International Conference on Systems, Man, and Cybernetics, SMC 2016—Conference Proceedings, pp. 1951–1956 (2017)
12. Kuli, D., Croft, E.A.: Estimating intent for human-robot interaction. In: IEEE International Conference on Advanced Robotics, pp. 810–815 (2003)

13. Lee, D., Ott, C.: Incremental kinesthetic teaching of motion primitives using the motion refinement tube. Auton. Robots **31**(2–3), 115–131 (2011)
14. Li, Y., Ge, S.S.: Human-robot collaboration based on motion intention estimation. IEEE/ASME Trans. Mechatron. **19**(3), 1007–1014 (2014)
15. Li, Y., Tee, K.P., Chan, W.L., Yan, R., Chua, Y., Limbu, D.K.: Continuous role adaptation for human robot shared control. IEEE Trans. Robot. **31**(3), 672–681 (2015)
16. Li, Y., Tee, K.P., Ge, S.S.: Switchings Between Trajectory Tracking and Force Minimization in Human–Robot Collaboration (HRC), pp. 65–81. Springer International Publishing, Cham (2017)
17. Lippiello, V., Cacace, J., Santamaria-Navarro, A., Andrade-Cetto, J., Trujillo, M., Esteves, Y.R., Viguria, A.: Hybrid visual servoing with hierarchical task composition for aerial manipulation. IEEE Robot. Autom. Lett. **1**(1), 259–266 (2016)
18. Mainprice, J., Berenson, D.: Human-robot collaborative manipulation planning using early prediction of human motion. In: 2013 IEEE/RSJ International Conference on Intelligent Robots and Systems (IROS), pp. 299–306 (2013)
19. Park, J.S., Park, C., Manocha, D.: Intention-aware motion planning using learning based human motion prediction (2016). CoRR arxiv:abs/1608.04837
20. Peternel, L., Babic, J.: Learning of compliant human-robot interaction using full-body haptic interface. Adv. Robot. **27**, 1003–1012 (2013)
21. Peternel, L., Tsagarakis, N., Caldwell, D., Ajoudani, A.: Adaptation of robot physical behaviour to human fatigue in human-robot co-manipulation. In: IEEE-RAS International Conference on Humanoid Robots, pp. 489–494 (2016)

Multi-modal Intention Prediction with Probabilistic Movement Primitives

Oriane Dermy, Francois Charpillet and Serena Ivaldi

Abstract This paper proposes a method for multi-modal prediction of intention based on a probabilistic description of movement primitives and goals. We target dyadic interaction between a human and a robot in a collaborative scenario. The robot acquires multi-modal models of collaborative action primitives containing gaze cues from the human partner and kinetic information about the manipulation primitives of its arm. We show that if the partner guides the robot with the gaze cue, the robot recognizes the intended action primitive even in the case of ambiguous actions. Furthermore, this prior knowledge acquired by gaze greatly improves the prediction of the future intended trajectory during a physical interaction. Results with the humanoid iCub are presented and discussed.

Keywords Multi-modality · Probabilistic movement primitive · Human robot interaction · Collaboration

1 Introduction

Humans are very good at mutually predicting and adapting their actions when collaborating with each other. Notably, they use multi-modal cues (acoustic, visual, etc.) to predict the intention of their partner in a robust way [25].

To collaborate proficiently with humans exhibiting anticipatory skills, robots also need to be able to predict the intention of human partners. Predicting the intention from a motion implies legibility and predictability, i.e., the robot must be able to

O. Dermy (✉) · F. Charpillet · S. Ivaldi
INRIA, 615 Rue du Jardin botanique, 54600 Villers-lès-Nancy, France
e-mail: oriane.dermy@inria.fr; oriane.dermy@gmail.com

F. Charpillet
e-mail: francois.charpillet@inria.fr

S. Ivaldi
e-mail: serena.ivaldi@inria.fr

© Springer International Publishing AG, part of Springer Nature 2019
F. Ficuciello et al. (eds.), *Human Friendly Robotics*, Springer Proceedings in Advanced Robotics 7, https://doi.org/10.1007/978-3-319-89327-3_14

quickly infer its goal and the future trajectory. Here, we advocate that the robot's prediction abilities can be improved by using multi-modal information [8, 27].

In our previous work [7], we addressed the problem of predicting the future intended trajectory during a physical human-robot interaction when the human partner moves the robot's arm to start a movement. We proposed to use Probabilistic Movement Primitives (ProMPs [21]) to learn the movement primitives from a set of demonstrations and to compute the intended trajectory given early observations of the action, guided by the human partner.

In this paper, both visual and kinetics cues are used to predict the human intent. The intention is modeled as a goal location and a trajectory that the robot has to perform with its arm. Both the robot's arm manipulations and the partner's gaze motions are learned as a multi-modal ProMP, that captures the distributions over the demonstrated trajectories.

From the physical inference, the robot is able to repeat movements and to continue movements initiated by the partner, even with few early-observations. From the visual inference the robot can predict and perform tasks that do not require the partner's guidance to refine the expected trajectory, but most importantly it can disambiguate easily among similar primitives.

The paper is organized as follows in Sect. 2. We briefly report on the literature about intention prediction and gaze as a conveyor of intention information. Section 3 formulates the problem settled in this paper. Section 4.1 summarizes the theoretical basis of the ProMP method to learn movement primitives, applied to learning multi-modal information. Section 5 presents a multi-modal intention recognition application, where results about the action recognition improve the prediction of the future trajectory. Finally, Sect. 6 discusses the proposed approach, its limitations and outlines our future developments.

2 Related Works

In order for the robot to predict the trajectory to be performed, it has to infer the user intention. Here, we focus on the inference from physical and visual cues. The paragraphs below provide a brief review of research literature on *intention* and *gaze* prediction. For the state of the art on *movement primitives* and *inference during pHRI*, we refer to [7].

Intention: Predicting the intention of a human essentially means predicting the goal of his/her current or upcoming action as well as the movement performed to reach this goal. Intention prediction is not only relevant to understand the prediction of intent between humans [5, 19], but also to allow robots to be understood by humans [10, 16], or to allow robots to understand humans in diverse applications, like human-robot collaboration [11, 26], and robot navigation [20]. Here, the gaze is used as a major cue to determine the user intention, coupling the directed gaze of the human with their associated actions.

Gaze as a conveyor of intention information: Directional gaze is the most fundamental cue for social interaction, as it enables mutual and joint attention. Hence, many studies consider the human face or gaze direction to interact with him. Some use this direction to estimate the user engagement with a robot companion [1, 6, 15]; or the user emotions to correct the robot's behavior [4]. Others to improve the robot behavior by ensuring the safety of the interaction [24]; by anticipating the action of their partner [13]; or by adapting robot actions according to the intention of their partner [17]. This last case corresponds to our current objective.

To complete this objective, the facial orientation or the human's gaze is first computed. Different methods are used to answer this question such as, Neural Networks [3], gradients computation [23], or probability. Gaze is often used as an a priori to perform an intended task (e.g., our work with ProMPs) to detect the object of interest (e.g., [12] with Neural Networks), or to predict the goal location (e.g., [22] with dynamic models). The main difference between our study and [12, 22] is that these works are interested in the human motion prediction while we associate human gaze to the robot motions.

In some research studies, the human's gaze direction is accurately measured using eye tracker [14, 20]. In our case, we rely on visual processing of the robot's cameras, which is less invasive and it does not require to wear a device, even though it is less accurate than eye tracker.

3 Problem Formulation

This paper proposes a method for multi-modal prediction of intention based on a probabilistic description of movement primitives and goals. We target dyadic interaction between a human and a robot, equipped with eyes and arms, in a pick and place collaborative scenario, shown in Fig. 1. In this scenario, different objects must be sorted following different trajectories. The human partner chooses to use visual

(a) Visual Guidance.   (b) Physical Guidance.

Fig. 1 The humanoid robot iCub **a** recognizes the intended movement primitive using the partner's directional gaze; **b** predicts the movement to perform using the partner's physical guidance at the beginning of the movement

and/or physical guidance to communicate the intended movement to the robot, that should be able at some point to continue the movement on its own. During the visual guidance, the robot tracks the partner's head orientation to predict his/her intention: the gaze trajectory is recognized as belonging to one of the known action primitives. The robot predicts then the current task and the future intended movement. It completes the intended task by placing the object in the expected place, following the trajectory intended by the partner. During physical guidance, the user starts to physically move the robot to perform the action; after early observations, the robot predicts the future movement to perform. If the human partner uses both modalities, the movement primitive can be recognized from the visual guidance (prior) and physical guidance can be used to refine the predicted trajectory (posterior). To realize this scenario, we make several hypotheses. Tracking the gaze using the eyes direction is difficult because of saccadic eye movement directed towards the goal, that could cause the gaze trajectory to be inconsistent. Therefore, the partner's head orientation is used to determine his intent. We assume the user's position with respect to the robot is almost fixed during the learning and the recognition task, because the robot learning is dependent on the partner's head orientation. We assume that the partner's head orientations when he/she looks at a same goal follow a normal distribution.

A conceptual representation of the problem is shown in Fig. 2. To learn the movement primitives (top), two partners run several demonstrations: one moves the robot's arm while another moves his head, following the trajectories to learn. From these demonstrations, the robot collects the Cartesian position of its arm and the partner's

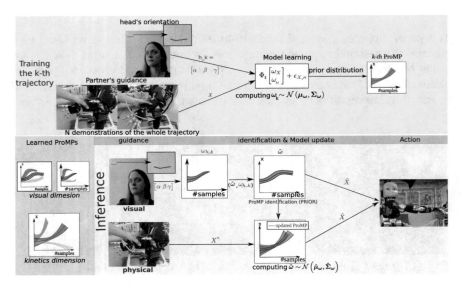

Fig. 2 Conceptual use of ProMP for predicting the desired trajectory to be performed by the robot. In the training phase (top), ProMPs are learned from several human demonstrations. In the inference phase (bottom), the robot recognizes the current ProMP using visual and/or physical information

gaze (head orientation). The trajectories make the base for learning the primitives (prior distribution). The bottom of the figure represents the inference step. The partner follows with his/her head the robot's movement and/or he/she physically initiates the robot's hand movement. When the prediction is done, the robot finishes autonomously the movement (i.e., drop the hand-held object). To show the improvement with respect to our previous work, the learned trajectories of the dropping phase have identical initial and final positions (making the prediction from early observations harder, and possible here only thanks to the multimodal primitive).

4 Methods

This section presents the ProMP method used to learn the motion primitives and to predict the trajectory of the ProMP given one modality. See [7] for further information.

4.1 Learning Motion Primitives with ProMP

A ProMP is a Bayesian parametric model of demonstrated trajectories in the form:

$$\xi(t) = \Phi_t \boldsymbol{\omega} + \epsilon_\xi \tag{1}$$

where $\xi(t)$ is the multimodal vector containing all the multi-modal variables to be learned at time t (e.g., $A(t) \in \Re^3$ for visual modality or $X(t) \in \Re^6$ for physical modality); $\boldsymbol{\omega} \in R^M$ is a time-independent parameter vector weighting the Φ matrix; $\epsilon_\xi \sim \mathcal{N}(0, \beta)$ is the trajectory noise; and Φ_t is a matrix of M Radial Basis Functions (RBFs) evaluated at time t: $\Phi_t = [\psi_1(t), \psi_2(t), \dots, \psi_M(t)]$. Note that all the ψ functions are scattered across time. The robot first records a set of n_1 trajectories $\{\Xi_1, \dots, \Xi_{n_1}\}$, where the i-th trajectory is $\Xi_i = \{\xi(1), \dots, \xi(t_{f_i})\}$. The duration t_{f_i} of each recorded trajectory varies, following the user demonstrations. To find a common representation (in terms of primitives), a time modulation is applied to all trajectories, such that they have the same number of samples $\bar{s}$. To do so, we consider "$\Phi_{\alpha t}$" instead of "Φ_t", to rescale the RBFs to each trajectory, with the time modulation parameter "$\alpha = \frac{\bar{s}}{t_{fi}}$". Such modulated trajectories are then used to learn a ProMP.

For each Ξ_i trajectory, we compute the ω_i parameter vector that minimizes the error between the observed $\xi_i(t)$ trajectory and its model $\Phi_{\alpha t}\omega_i + \epsilon_\xi$. This is done using the Regularized Least Mean Square algorithm.

Thus, we obtain a set of parameters upon which a normal distribution is computed:

$$p(\boldsymbol{\omega}) \sim \mathcal{N}(\mu_\omega, \Sigma_\omega) \tag{2}$$

$$\text{with } \mu_{\omega} = \frac{1}{n} \sum_{i=1}^{n} \omega_i \tag{3}$$

$$\text{and } \Sigma_{\omega} = \frac{1}{n-1} \sum_{i=1}^{n} (\omega_i - \mu_{\omega})^{\top} (\omega_i - \mu_{\omega}) \tag{4}$$

4.2   Predicting the Trajectory of the ProMP

The learned ProMPs corresponds to several skills or action primitives. They are used as a prior knowledge by the robot to predict the current action and its future trajectory, so that it can continue the movement autonomously. Here, early observations of the trajectory are a subset of the variables to learn:

$$\Xi^o = [\Xi_1, \dots, \Xi_{n_o}]^{\top} = \{X^o || A^o || \begin{bmatrix} X^o \\ A^o \end{bmatrix}\} \tag{5}$$

where X^o is the haptic measurement and A^o, the visual measurement.

The first step of the recognition process is to recognize the current ProMP $\hat{k} \in [1 : 2]$, and the temporal modulation parameter $\hat{\alpha}$ from this partial observation Ξ^o. This is done by computing the most likely couple of temporal modulation parameter and ProMP type $(\hat{\alpha}_{\hat{k}}, \hat{k})$ corresponding to the early trajectory. We use two methods to perform this computation.

- The first called "*maximum likelihood*" (*ML*) is computed by:

$$(\hat{\alpha}_{\hat{k}}, \hat{k}) = \text{argmax}_{(\alpha \in S_{\alpha_{\hat{k}}}, \hat{k} \in [1:2])} \{\text{loglikelihood}(\Xi^o, \mu_{\omega_{\hat{k}}}, \sigma_{\omega_{\hat{k}}}, \alpha_{\hat{k}})\} \tag{6}$$

  where $S_{\alpha_{\hat{k}}} = \{\alpha_{\hat{k}1}, \dots, \alpha_{\hat{k}n}\}$ is the set of all the α parameters computed during the learning for each observation of the ProMP $\hat{k}$.
- The second called "*model*" is based on the assumption there is a correlation between the time modulation α and the variation of the trajectory δ_{n_o} from the beginning until the instant n_o. Indeed, we assume that the time modulation parameter α is linked to the movement speed, which can be roughly approximated by "$\dot{\Xi} = \frac{\delta \Xi}{t_f}$". For the physical inference, the "variation" of the hand position is computed by "$\delta_{n_o} = X(n_o) - X(1)$", whereas for the visual inference, the variation of the partner's head orientation is computed by "$\delta_{n_o} = A(n_o) - A(1)$". We model the mapping between δ_{n_o} and α by:

$$\alpha = \Psi(\delta_{n_o})^{\top} \omega_{\alpha} + \epsilon_{\alpha}, \tag{7}$$

where Ψ are RBFs, and ϵ_{α} is a zero-mean Gaussian noise. During learning, we compute the ω_{α} parameter, using the same method as in Eq. 1 and during the inference, we compute $\hat{\alpha} = \Psi(\delta_{n_o})^\top \omega_\alpha$. Finally, we compute the maximum likelihood in the set of $\{\hat{\alpha_1}, \hat{\alpha_2}\}$.

Once identified the $(\hat{\alpha}_{\hat{k}}, \hat{k})$ couple, the recognized distribution (called the "prior") can be updated by:

$$
\begin{cases}
\hat{\mu}_{\omega_{\hat{k}}} = \mu_{\omega_{\hat{k}}} + K(\Xi^o - \Phi_{\hat{\alpha}_{\hat{k}}[1:n_o]}\mu_{\omega_{\hat{k}}}) \\
\hat{\Sigma}_{\omega_{\hat{k}}} = \Sigma_{\omega_{\hat{k}}} - K(\Phi_{\hat{\alpha}_{\hat{k}}[1:n_o]}\Sigma_{\omega_{\hat{k}}}) \\
K = \Sigma_{\omega_{\hat{k}}} \Phi_{\hat{\alpha}_{\hat{k}}[1:n_o]}^\top (\Sigma_{\xi^o} + \Phi_{\hat{\alpha}_{\hat{k}}[1:n_o]}\Sigma_{\omega_{\hat{k}}} \Phi_{\hat{\alpha}_{\hat{k}}[1:n_o]}^\top)^{-1}
\end{cases}
\tag{8}
$$

with $\hat{\alpha}_{\hat{k}}[1:n_o] = \hat{\alpha}_{\hat{k}}\, t$ (in matrix form), with $t \in [1:n_o]$. This is the posterior distribution.

Finally, the inferred trajectory is given by:

$$
\forall t \in [1:\hat{t}_f], \hat{\xi}(t) = \Phi_t\, \hat{\mu}_{\omega_{\hat{k}}}
$$

with the expected duration of the trajectory $\hat{t}_f = \frac{\bar{s}}{\hat{\alpha}_k}$. The robot is now able to finish the movement executing the most-likely "future" trajectory $\hat{X} = [\hat{X}_{n_o+1}, \ldots, \hat{X}_{\hat{t}_f}]^\top$.

5 Experiments

5.1 Experimental Setup

We carried out experiments with the humanoid robot iCub. To retrieve the approximated gaze direction, we use the roll/pitch/yaw angles of the user's head orientation, extracted from the camera image of the iCubs eyes by Intraface [28], To retrieve the Cartesian information, we use an iCub module that computes the Cartesian position and orientation (iKinCartesianSolver). The experimental procedure is outlined in Fig. 2. The training phase requires a human partner to manipulate the robot's left arm to perform kinesthetic teaching and a second partner to guide the robot via gaze. In the inference phase, only the partner interacts with the robot.

5.2 Teaching iCub the Action Primitives

We taught the robot two multi-modal movement primitives that make it drop an object inside a target bin (roughly at the same position) but following two different type of trajectories coupled with the corresponding trajectories of the human partner. These primitives contain the Cartesian position and orientation of the robot's left hand (guided by the robot operator), and the head orientation of the human partner

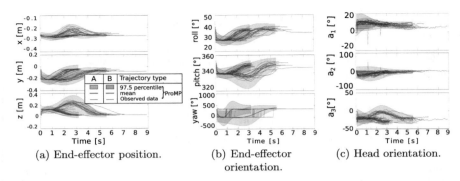

(a) End-effector position.   (b) End-effector   (c) Head orientation.
                                        orientation.

Fig. 3 Demonstrations (trajectories) and primitives. In red (ProMP A) the "curved" trajectory, and in blue (ProMP B) the "direct" trajectory

that visually guides the robot: $\xi(t) = [X(t), A(t)]^{\top}$, with $X(t) \in \Re^6$ the Cartesian pose and $A(t) \Re^3$ the roll-pitch-yaw orientation angles of the partner's head.

We performed 20 trajectory demonstrations per primitive action. Figure 3 shows the demonstrations and the learned-distribution for the two ProMPs.

5.3 Activating Primitives with Gaze

The gaze cue is used to identify the current action. This procedure has two advantages. First, it does not require physical interaction, which could ease interacting with the robot for some people. Second, it enables to improve the prediction of intended trajectory, especially in case of ambiguous primitives that overlap and could make it difficult to obtain a good prediction with few early observations. An intuitive case is shown in Fig. 4.

From [7], we retain two methods to compute the time modulation: "maximum likelihood" (*ML*) and "*model*", where the latter consists on estimating the trajectory duration according to the global partner's head orientation variation: "$\delta_{n_o} = A(n_o) - A(1)$".

We tested off-line the gaze prediction of the trajectories on the acquired data set using cross-validation. Figure 5 shows a prediction example after having observed 50% of the trajectory. The inferred trajectory is the mean trajectory of the red posterior distribution. Note that this posterior distribution is included in the prior distribution and pass by the observed data with some *flexibility*, that correspond to the expected measurement noise fixed a-priori. Even though the partner's head orientation observations are not accurate, the prediction is good enough to allow the robot to complete the task correctly.

Figure 6a represents the error of ProMP recognition according to the percentage of observations of the test trajectory. The longer the head trajectory is observed, the smaller is the prediction error, for both methods for computing the time modulation.

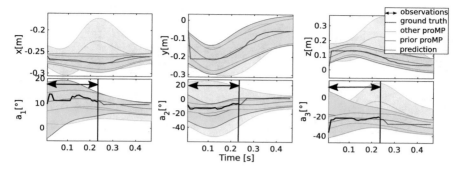

Fig. 4 Example of position inference from 50% of the head orientation trajectory. The dots represent the trajectory the robot has to perform (ground truth). The black curves represent the measurements done by the robot. The blue distribution represents the recognized ProMP and the green distribution the other ProMP. The red distribution represents the posterior of the blue distribution, computed from the measured data

Fig. 5 Gaze helps disambiguate two overlapping primitives

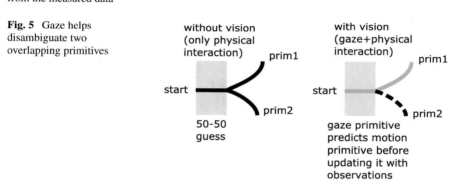

This figure also shows that the *model* is less accurate than the *ML* method when the robot observes less than 70% of the whole trajectory, while with more observation the *model* method is a slightly more accurate. Since head movements are fast, the robot can use the whole head movement trajectory and still react quickly. So, we can use the *model* method to allow the robot to recognize which ProMP to follow for the visual guidance. With 70% observation of a trajectory, there is no ProMP type recognition error, thus, the robot can roughly infer the trajectory to perform (which corresponds to 3 s).

We represent in Fig. 6b the average error of the Cartesian position of the inferred trajectory. It shows that the error of the predicted trajectory goes from 4 cm (When the robot observes 10% of the trajectory) to 2 cm (80%). Thus, the more the robot observes its partner's head trajectory, the more it is able to achieve it own movement intended by its partner.

However, we can wonder if the posterior distribution is more accurate than the prior. It would be the case if the partner's head orientation was totally correlated to the robot's hand position and the measurement accurate enough to infer exactly the end-

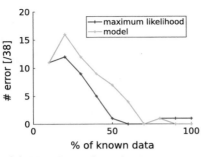 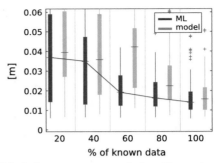

(a) Number of prediction errors.   (b) Inference error of the Cartesian position: average$|X_{des} - \hat{X}|$.

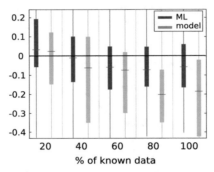

(c) RNMSE difference between the prior and posterior distribution.

Fig. 6 Visual guidance analysis

trajectory. Figure 6c represents the difference of the Normalized Root Mean Square Error (NRMSE) between the prior and the posterior distribution. From 40% of the trajectory observation, this difference is inferior to zero, meaning that by updating the distribution, the robot improves the trajectory inference. Thus, the visual guidance can be used to determine which ProMP the robot has to follow, but also to adapt the ProMP distribution from the user's head guidance in an accurate way.

To achieve a better accuracy, we assume the physical interaction will more indicated. To verify this assumption, the next session presents the physical guidance experiment.

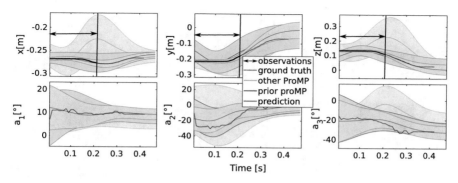

Fig. 7 Example of trajectory inference from physical guidance

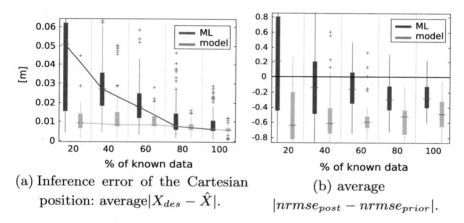

(a) Inference error of the Cartesian position: average$|X_{des} - \hat{X}|$.

(b) average $|nrmse_{post} - nrmse_{prior}|$.

Fig. 8 Physical guidance analysis

5.4 Inference of Intended Trajectories with Physical Guidance

The same prediction experiment from early-demonstrations than the previous section is presented here with haptic signals. Figure 7 presents an example of such prediction. If we compare to the visual experiment, we can note that the inferred trajectory (mean of the red posterior distribution) is closer to the ground truth. Figure 8a verifies this idea. It represents the average distance between the inferred trajectory ($\hat{X}$) and the ground truth (X_{des}), and the results show that the trajectory prediction using physical estimation is more accurate than the visual estimation, whether with the *model* or the *ML* method, with an average of less than 1 cm of distance error for the *model* and from 3 cm (40% of known data) to 1 cm (80%) for the *ML*. Moreover, Fig. 8b shows that the posterior distribution of the ProMP improves the accuracy of the trajectory, mainly for the *model* method which explains why the distance error using this method is short in the previous figure.

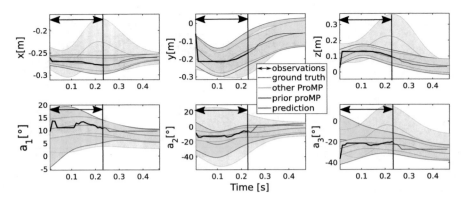

Fig. 9 Example of position inference from 50% of the head orientation and the Cartesian position trajectories

Now, we can wonder if using the two modalities could improve the performance of this inference ability. Thus, the next section is the multi-modal experiment on the same data set.

5.5 Inference of Intended Trajectories with Multi-modal Guidance

Figure 9 represents the inference of the Cartesian position trajectory when the robot knows 50% of the trajectory data to achieve and when it uses both visual and physical measurements (black curves). In this example, the inferred trajectory (mean of the red posterior distribution) is close to the trajectory expected by the partner (black dots). To compare this multi-modal prediction with visual or physical prediction only, Figs. 10 and 11 represent statistics for each prediction type. Figure 10 represents the distance error between the Cartesian position of the expected and the inferred trajectory. Whether with the *model* (in Fig. 10a) or the *ML* method (in Fig. 10b), the inference using the Cartesian position measurement only is more accurate than using the multi-modal or the visual-only measurement. The performance of this physical guidance is mainly visible with the *model* method, where the distance error is really short. Thus, the multi-modality guidance did not improve the inference ability of the robot.

From Fig. 11, we can see the number of ProMP recognition error according to the type of modality used to perform the inference. An interesting result is that by using the *model* method (in Fig. 11a), the robot is entirely able to recognize the initiated movement from 70% of know data, and with the *ML* method, the robot has only done one error from the 38 trials (which corresponds to 2%). Thus, the multi-modal clearly improves the ProMP recognition step of the inference, even though it did not improve the final inferred trajectory precision.

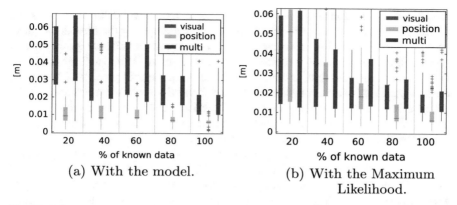

(a) With the model.

(b) With the Maximum Likelihood.

Fig. 10 Inference error of the Cartesian position: average $|X_{des} - \hat{X}|$ according to modality used

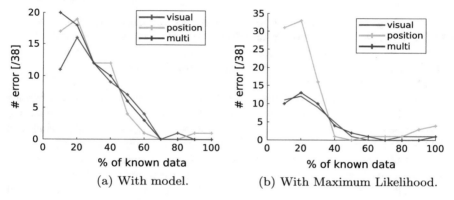

(a) With model.

(b) With Maximum Likelihood.

Fig. 11 Prediction error according to modality used

6 Conclusions

This paper presents a multi-modal method for robots to predict the partner's intended trajectory during HRI using haptic and/or gaze cues. We tested our system with the humanoid iCub collaborating with a human partner in a task where the robot has to grasp an object using different trajectories. The human physically interacts with the robot's left arm to start an action and/or uses his directional gaze to guide the robot. We build on our previous work [7], where elementary actions are represented by Probabilistic Movement Primitives that enable prediction of goals from early observations. During physical guidance, the robot uses the haptic information to recognize the current action, then it is able to accurately predict the goal, the future intended trajectory and its duration. A limitation of previous inference method is that the robot is not able to determine which movement primitive to follow when the early-observations are ambiguous, i.e., identical to more than one primitive. In that case, the visual guidance is used to identify the correct movement primitive. While during the visual guidance, the same prediction is done using the directional gaze, approximated here by the head orientation. The association between gaze cues and robot primitives is done by a multi-modal learning phase. The visual modality has two main advantages: first, it does not require the partner to physically touch the robot to start his intended movement; second, it provides a faster recognition of the action primitive if compared with physical signals. However, results show that by using the visual instead of the physical guidance, the performance of the inference decreases slightly (around 1.5 cm). A limit of this modality is the accuracy of the gaze estimation. To improve it, we have many possibilities: use the Kinect to have more relevant data; use another head recognition software instead of Intraface; or use the Xsens 3D tracking. It is also possible to add another "*no-human*" modality to even surpass human inference skills, by guiding the robot from a watch that contains sensors to detect the human partner's arm pose and to use this pose to learn and recognize ProMPs.

Regarding the inference using multi-modal measurements, results show that by adding the visual recognition in addition to the physical recognition, it did not improve the accuracy of the inferred trajectory (i.e., it did not improve the posterior distribution computation), but it improves the ProMP recognition (i.e., it improves the first step of the inference that consists on recognizing which movement the robot has to execute among the one it has learned). Thus, to have the better inference skills, we should use the multi-modal guidance to allow robots to recognize the movement/action to perform, and then we should use the haptic guidance to improve the movement precision according to the early measurements. However, the multi-modal guidance currently requires to use two human partners (one in front of the robot to guide it with his/her head and the other one to guide it physically) or to perform the guidance type one after the other. The utilization of the Xsens is a good way to improve this study because one partner will be able to guide physically and visually the partner at the same time, hence in a more natural way.

In future work, we will also study the human preference for the use between the haptic and visual guidance modes.

Acknowledgements The authors wish to thank Olivier Rochel, Alexandros Paraschos, Marco Ewerton, Waldez Azevedo Gomes Junior and Pauline Maurice for their help and feedbacks.

References

1. Anzalone, S.M., Boucenna, S., Ivaldi, S., Chetouani, M.: Evaluating the engagement with social robots. Int. J. Soc. Robot. **7**(4), 465–478 (2015)
2. Bader, T., Vogelgesang, M., Klaus, E.: Multimodal integration of natural gaze behavior for intention recognition during object manipulation. In: Proceedings of PIC on Multimodal Interfaces, pp. 199–206. ACM (2009)
3. Baluja, S., Pomerleau, D.: Non-intrusive gaze tracking using artificial neural networks. In: Proceedings of Advances in NIPS, pp. 753–760 (1994)

4. Boucenna, S., Gaussier, P., Andry, P., Hafemeister, L.: A robot learns the facial expressions recognition and face/non-face discrimination through an imitation game. Int. J. Soc. Robot. **6**(4), 633–652 (2014)
5. Bretherton, I.: Intentional communication and the development of an understanding of mind. Children's Theories of Mind: Mental States and Social Understanding, pp. 49–75 (1991)
6. Castellano, G., Pereira, A., Leite, I., Paiva, A., McOwan, P.W.: Detecting user engagement with a robot companion using task and social interaction-based features. In: Proceedings of PIC on Multimodal Interfaces, pp. 119–126. ACM (2009)
7. Dermy, O., Paraschos, A., Ewerton, M., Peters, J., Charpillet, F., Ivaldi, S.: Prediction of intention during interaction with ICUB with probabilistic movement primitives. Front. Robot AI (2017)
8. Dillmann, R., Becher, R., Steinhaus, P.: ARMAR II-a learning and cooperative multimodal humanoid robot system. Int. J. Humanoid Robot **1**(01), 143–155 (2004)
9. Dragan, A., Srinivasa, S.: Generating legible motion. In: Proceedings of Robotics: Science and Systems. Berlin, Germany, June 2013
10. Dragan, A., Srinivasa, S.: Integrating human observer inferences into robot motion planning. Auton. Robot. **37**(4), 351–368 (2014)
11. Ferrer, G., Sanfeliu, A.: Bayesian human motion intentionality prediction in urban environments. Pattern Recogn. Lett. **44**, 134–140 (2014)
12. Hoffman, M.W., Grimes, D.B., Shon, A.P., Rao, R.P.: A probabilistic model of gaze imitation and shared attention. Neural Netw. **19**(3), 299–310 (2006)
13. Huang, C.M., Mutlu, B.: Anticipatory robot control for efficient human-robot collaboration. In: Proceedings of HRI, pp. 83–90 (2016)
14. Ishii, R., Shinohara, Y., Nakano, T., Nishida, T.: Combining multiple types of eye-gaze information to predict user's conversational engagement. In: 2nd Workshop on Eye Gaze on Intelligent Human Machine Interaction (2011)
15. Ivaldi, S., Lefort, S., Peters, J., Chetouani, M., Provasi, J., Zibetti, E.: Towards engagement models that consider individual factors in HRI. Int. J. of Soc. Robot. **9**, 63–86 (2017)
16. Kim, J., Banks, C.J., Shah, J.A.: Collaborative planning with encoding of users' high-level strategies. In: Proceedings of AAAI (2017)
17. Kozima, H., Yano, H.: A robot that learns to communicate with human caregivers. In: Proceedings of the First International Workshop on Epigenetic Robotics, pp. 47–52 (2001)
18. Ma, C., Prendinger, H., Ishizuka, M.: Eye movement as an indicator of users' involvement with embodied interfaces at the low level. In: Proceedings of AISB, pp. 136–143 (2005)
19. Meltzoff, A.N., Brooks, R.: Eyes wide shut: the importance of eyes in infant gaze following and understanding other minds. In: Flom, R., Lee, K., Muir, D. (eds.) Gaze Following: Its Development and Significance. Erlbaum. [EVH] (2007)
20. Mitsugami, I., Ukita, N., Kidode, M.: Robot navigation by eye pointing. Lecture notes in computer science **3711**, 256 (2005)
21. Paraschos, A., Daniel, C., Peters, J.R., Neumann, G.: Probabilistic movement primitives. In: Proceedings of NIPS, pp. 2616–2624 (2013)
22. Ravichandar, H., Kumar, A., Dani, A.: Bayesian human intention inference through multiple model filtering with gaze-based priors. In: Proceedings of Information Fusion (FUSION), pp. 2296–2302. IEEE (2016)
23. Timm, F., Barth, E.: Accurate eye centre localisation by means of gradients. In: Proceedings of Visapp, vol. 11, pp. 125–130 (2011)
24. Traver, V.J., del Pobil, A.P., Pérez-Francisco, M.: Making service robots human-safe. In: Proceedings of (IROS 2000), vol. 1, pp. 696–701. IEEE (2000)
25. Walker-Andrews, A.S.: Infants' perception of expressive behaviors: differentiation of multimodal information. Psychol. Bull. **121**(3), 437 (1997)
26. Wang, Z., Deisenroth, M.P., Amor, H.B., Vogt, D., Schölkopf, B., Peters, J.: Probabilistic modeling of human movements for intention inference. Robot. Sci. Syst. (2012)

27. Weser, M., Westhoff, D., Huser, M., Zhang, J.: Multimodal people tracking and trajectory prediction based on learned generalized motion patterns. In: International Conference on Multisensor Fusion and Integration for Intelligent Systems, pp. 541–546 (2006)
28. Xiong, X., De la Torre, F.: Supervised descent method and its applications to face alignment. In: Proceedings of IEEE CVPR (2013)

Locomotion Mode Classification Based on Support Vector Machines and Hip Joint Angles: A Feasibility Study for Applications in Wearable Robotics

Vito Papapicco, Andrea Parri, Elena Martini,
Vitoantonio Bevilacqua, Simona Crea and Nicola Vitiello

Abstract Intention decoding of locomotion-related activities covers an essential role in the control architecture of active orthotic devices for gait assistance. This work presents a subject-independent classification method, based on support vector machines, for the identification of locomotion-related activities, i.e. overground walking, ascending and descending stairs. The algorithm uses features extracted only from hip angles measured by joint encoders integrated on a lower-limb active orthosis for gait assistance. Different sets of features are tested in order to identify the configuration with better performance. The highest success rate (i.e. 99% of correct classification) is achieved using the maximum number of features, namely seven features. In future works the algorithm based on the identified set of features will be implemented on the real-time controller of the active pelvis orthosis and tested in activities of daily life.

V. Papapicco (✉) · A. Parri · E. Martini · S. Crea · N. Vitiello
The BioRobotics Institute, Scuola Superiore SantAnna, Pontedera, Italy
e-mail: vito.papapicco@santannapisa.it

A. Parri
e-mail: andrea1.parri@santannapisa.it

E. Martini
e-mail: elena.martini@santannapisa.it

S. Crea
e-mail: simona.crea@santannapisa.it

N. Vitiello
e-mail: nicola.vitiello@santannapisa.it

N. Vitiello
Don Carlo Gnocchi Foundation, Milan, Italy

V. Bevilacqua
Department of Electrical and Information Engineering (DEI),
Polytechnic University of Bari, Bari, Italy
e-mail: vitoantonio.bevilacqua@poliba.it

© Springer International Publishing AG, part of Springer Nature 2019
F. Ficuciello et al. (eds.), *Human Friendly Robotics*, Springer Proceedings
in Advanced Robotics 7, https://doi.org/10.1007/978-3-319-89327-3_15

197

1 Introduction

Lower-limb impairments are often consequences of age-related neurological or cardiovascular diseases, such as stroke or diabetes [1], and due to the increment of life expectancy [2], their occurrence is foreseen to increase in next future. A sustainable aging population requires the development of new solutions to maintain elderly active in daily life. Exoskeletons are an example of assistive devices conceived to improve the recovery or improve mobility in daily life of people affected by lower-limb impairments [3]. Exoskeletons are mechatronic devices that are worn by the user and operate in parallel with his/her limbs, in a close physical and cognitive cooperation. Despite the huge potential in improving motor capabilities of the end-user [4], there are still open technological challenges. In order to assist the user safely and reliably, the control system of the wearable robot must be able to detect his/her movement intention, dealing robustly with intra- and inter-subject variability [5].

In the last years the research has shown interest in the development of control strategies that are compliant with the residual motor skills of the user, exploiting a wide variety of classification techniques and set of sensors [5, 6]. Classification methods typically use electromiographic (EMG) signals [7] and/or mechanical sensors such as load cells, encoders, or inertial measurement units (IMUs). Sensors can be embedded in the mechatronic architecture of the wearable robot [8–11] or can be worn in addition to the robot [12–14]. Clearly, minimizing the number of additional sensors would improve the acceptability of the system by the user but might influence the accuracy rate of the classification.

Performances of intention detection methods depend also on the techniques used to manipulate and evaluate sensory data. Very high accuracy rates have been reached using supervised and unsupervised machine learning methods, fuzzy logic strategies or finite state machines [5]. Some of these methods require subject-specific training or tweaks in order to operate, and some are not adaptive to changes of the gait speed or to different gait patterns.

This work presents a subject-independent classification method based on support vector machines (SVM) for the classification of locomotion-related activities, i.e. overground walking, ascending and descending stairs. The algorithm uses features extracted from hip joint flexion/extension angles and runs independently on left-leg and right-leg data; kinematic data have been acquired wearing a hip exoskeleton, namely the Active Pelvis Orthosis (APO), controlled in both transparent and assistive mode [15] and processed offline. In transparent mode the desired torque of each joint is set to zero, while in assistive mode the desired torque is calculated depending on the gait phase and the locomotion mode, in order to actively help the wearer in the flexion and extension of the hip. Five different sets of features are extracted and used to train and test the algorithm, with the final goal of identifying the optimal set of features to achieve accuracy rate higher than 99%.

2 Materials and Methods

2.1 Active Pelvis Orthosis

The APO is a wearable robotic assistive device conceived to assist hip flexion-extension movements during locomotion-related activities. The APO used in this work is an advanced light-weight portable version of the tethered device presented in [15]. Its mechanical structure consists of a main carbon fibre plate connecting the exoskeleton to an orthopedic cuff enveloping the users trunk. Two carbon-fibre lateral extensible arms are endowed with two degrees of freedom (DoFs) both collocated with their anatomical counterpart: a passive hip adduction-abduction DoF and an active hip flexion-extension DoF. Actuation units (one for each arm) are based on a series elastic actuation (SEA) architecture [16]. Each SEA has a single-axis configuration composed of a motor-reduction stage connected to a torsional spring, whose deformation is measured by an absolute encoder. A second encoder placed on the hip axis measures the hip angle (positive in flexion). A posterior backpack connected to the carbon fibre plate contains the control board, namely a NI sbRIO-9651 (National Instrument, Austin, Texas, US), electronics, motor drivers and batteries.

The APO control architecture has a hierarchical structure, with two layers, namely a low-level closed-loop torque control layer and a high-level layer. The high-level control layer runs at 100 Hz frequency on a real-time environment and implements the assistive strategies and the locomotion mode classification methods. Two low-level closed-loop torque controllers (one for each actuator) based on classical PD compensator are implemented on the FPGA, running at 1 kHz.

2.2 General Procedures

Sixteen healthy subjects (13 male and 3 female subjects, age: 27.3 ± 3.7; weight: 74.4 ± 11.8 kg; height: 1.73 ± 0.06 m) participated to this study.

Upon arrival subjects were requested to wear the exoskeleton and familiarize with the APO under transparent and assistive modes.

After familiarization, subjects performed different locomotion activities: stair ascending (SA), stair descending (SD) and ground-level walking (GL) (Fig. 1) at different self-selected speeds, under transparent and assistive mode conditions.

Measured hip joint angles were saved for offline analysis.

2.3 Offline Features Extraction and Creation of the Dataset

Left and right joint angles were offline segmented in stride cycles and labeled according to the performed locomotion-related activity.

Strides were segmented based on the detection of the maximum hip flexion angle and offline resampled over 100 samples.

(a)　　　　**(b)**

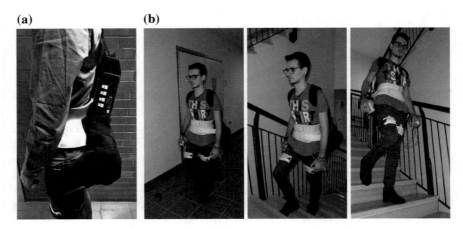

Fig. 1 **a** A close-up of lateral side of the APO. **b** One representative subject performing GL, SA and SD wearing the APO during the collection of stride samples

For each stride, the following features were extracted from relative leg joint angles and velocities (some of them are shown in Fig. 2), the latter calculated offline as the first-derivative of the joint angle:

- Maximum hip joint angle (HJA Max)
- Minimum hip joint angle (HJA Min)
- Range of motion (HJA RoM)
- Average hip joint angle (HJA Mean)
- Hip joint velocity at 19% of the stride, i.e. the foot contact (19% HJV)
- Percentage of the minimum hip joint angle (HJA Min Pos)
- Hip joint velocity at 50% of the stride (50% HJV).

Two separate datasets were created, one for the training and one for the testing.

Data of fourteen subjects were used to train the algorithm and data from the remaining two subjects for testing.

The training dataset consisted of 408 strides, i.e. 136 stride cycles for each locomotion mode. Majority of the strides were acquired in transparent mode (i.e. 312 strides), whereas few were acquired in assistive mode (i.e. 96 strides).

A total of 306 strides (i.e. 258 in transparent and 48 in assistive mode) were stored in the testing set, with 102 strides for each locomotion mode.

2.4 Algorithm

The algorithm developed in this work is based on SVMs. SVMs are a supervised machine learning method used for binary classification problem [17]. Due to the low computational complexity for real time applications and the high classification

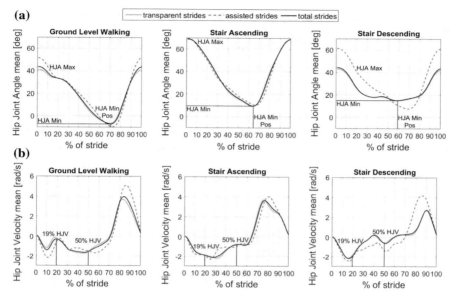

Fig. 2 Across-stride and across-subjects averaged hip joint angles (**a**) and velocities (**b**) obtained from the training dataset in the three locomotion modes for the strides under transparent mode (dotted line), the assisted strides (dashed line) and all the strides (full line). The features are reported on the plots of the different locomotion modes

performance, SVMs have been used in a variety of fields, including medical applications [18]. The training of the SVM has the goal to identify an hyper-plane that separates two classes in the space of the features, maximizing the margin between the boundary function and its closest points of each class.

In this study the classes were the three locomotion modes and the strides the points in the space, represented by the values of the features.

The five different SVM configurations differ in the number of features used, as Table 1 shows.

A direct acyclic graph (DAG) strategy allows to implement a multi-class classification, using a cascade of binomial classifiers [19]. For a N-class problem, the number of *decision nodes* is N(N − 1)/2; in this study, a three-nodes binary tree has been implemented (Fig. 3). Each decision node consisted of ten SVMs, devoted to perform an election process. At each node, the extracted features were used to exclude one class from the three possible classes. The SVMs of each node have been obtained performing a 10-fold cross-validation [20] over the training dataset.

The first decision node allowed to exclude one mode, between SA and SD: the SVMs of this decision node, indeed, had the higher minimum margin, with respect to the other binary combinations of classes. The other two nodes then classified the locomotion mode, based on the two possibilities left after the first decision node.

Table 1 Features selected for the different configurations of the classification method. The × marks the features used

	HJA Max	HJA Min	HJA RoM	HJA Mean	19% HJV	HJA Min Pos	50% HJV
#1	×	×	×				
#2	×	×	×	×			
#3	×	×	×	×	×		
#4	×	×	×	×	×	×	
#5	×	×	×	×	×	×	×

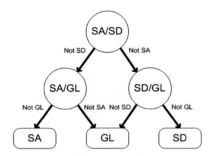

Fig. 3 DAG structure used for the locomotion mode classification problem

Performance of the classification algorithm configurations were quantified by the overall success rate and confusion matrices.

3 Results

Overall accuracy rate and the confusion matrices of the algorithm using the five sets of features are shown in Table 2.

The accuracy rate of the algorithm using 3 features resulted 78.80%, while performance increased with number of features up to 99.01% when using 7 features.

The classification accuracy of the SA modality resulted 100% with all the sets of features, while the highest classification accuracy of the SD modality was achieved using 7 features (i.e. 98%). For the GL the highest performance was achieved using 5 features (i.e. 100%), but good performance (i.e. 99%) were achieved also with a higher number of features.

The confusion matrix of the first configuration, namely the 3 features method, classifies 34% of GL strides as SA, and 27% of SD strides as SA. Using 4 features, the rate of GL strides labeled as SA decreases to 1%, while the rate of SD strides labeled as SA is 7%. This last rate is further reduced in the other configurations, decreasing to a minimum of 2% with the 7 features method.

Table 2 Overall accuracy rate of the classification for the different sets of features and confusion matrices for the different configurations, over the test dataset. Rows of confusion matrices are the actual stride classes and columns are the predicted classes. Therefore diagonal cells show the accuracy of each locomotion mode

Configuration	Overall Accuracy
#1	78.80%
#2	97.02%
#3	98.35%
#4	98.68%
#5	99.01%

#1	Detected Mode		
Actual Mode	GL	SD	SA
GL	64%	2%	34%
SD	0%	73%	27%
SA	0%	0%	100%

#2	Detected Mode		
Actual Mode	GL	SD	SA
GL	98%	1%	1%
SD	0%	93%	7%
SA	0%	0%	100%

#3	Detected Mode		
Actual Mode	GL	SD	SA
GL	100%	0%	0%
SD	0%	95%	5%
SA	0%	0%	100%

#4	Detected Mode		
Actual Mode	GL	SD	SA
GL	99%	0%	1%
SD	0%	97%	3%
SA	0%	0%	100%

#5	Detected Mode		
Actual Mode	GL	SD	SA
GL	99%	0%	1%
SD	0%	98%	2%
SA	0%	0%	100%

4 Discussion

The design of control strategies that can effectively identify the user movement intention is still an open challenge in the field of assistive wearable robots. Existing locomotion mode classifiers often integrate onboard sensors with additional EMG or mechanical sensors aiming at achieving successful classification. The goal of this work was to validate a subject-independent SVM-based classification algorithm, designed to identify three different locomotion-related activities, using only hip joint kinematics information measured from hip encoders of an APO.

The configuration with seven features showed the highest accuracy in the classification of the full stride cycle. In that configuration the overall accuracy rate and the accuracy rate in each locomotion mode resulted higher than 98%. The SA mode was classified correctly using all the configurations, in both transparent and assisted strides, meaning that it can be easily identified using the considered features.

It is worth noting that the proposed method can be easily implemented in online applications with limited computational burden for the APO real time controller in order to autonomously and consistently adapt the assistive action to the different locomotion tasks performed by the user: indeed, since all of the inspected SVMs had linear separation boundaries, the computational time to perform the classification did not appreciably increase with the number of features.

The overall accuracy of the method resulted comparable with the results of other studies regarding locomotion tasks recognition. A previous work from our group reported similar classification performance (i.e. 99.4% classification accuracy), using a previous version of the APO and pressure-sensitive insoles to provide additional gait information [21]. The method presented in this work improves the previous approach since it does not require sensor external to the ones integrated on the APO.

Moreover, in [22], Jang and colleagues presented a classification method based on fuzzy logic that uses hip joint angles and one IMU, embedded on a lower-limb exoskeleton. Strides were recorded with the exoskeleton in transparent mode, and the overall accuracy resulted 97.4%. Despite the good performance, the study did not consider strides acquired with the exoskeleton controlled in assistive mode.

The SVM-DAG-based algorithm presented in this work demonstrated a huge potential for the classification of different locomotion tasks, with independency from subjects and gait cadence. Notably, the strides collected in this study included transparent and assistive modes, demonstrating that the algorithm can be effectively applied when the system is active and the gait pattern is modified by the assistance.

Future tests will be carried out in order to validate the online performance of the recognition while performing locomotion-related activities wearing the APO. Subjects with lower-limb impairments will be recruited to assess the performance with abnormal gait patterns.

Acknowledgements This work was supported in part by the EU within the CYBERLEGs Plus Plus project (H2020-ICT-2016-1 Grant Agreement #731931) and in part by INAIL within the MOTU project (PPR-AI 1-2).
Andrea Parri, Simona Crea and Nicola Vitiello have commercial interests in IUVO s.r.l., a spin off company of Scuola Superiore SantAnna. Currently, the IP protecting the APO technology has been licensed to IUVO s.r.l. for commercial exploitation.

References

1. Verghese, J., Levalley, A., Hall, C.B., Katz, M.J., Ambrose, A.F., Lipton, R.B.: Epidemiology of gait disorders in community-residing older adults. J. Am. Geriatr. Soc. **54**, 255–261 (2006)
2. World Health Organization: Global Health and Aging (2006). http://www.who.int/ageing/publications/globalhealth.pdf
3. Pons, J.L.: Wearable Robots: Biomechatronic Exoskeletons. Wiley, Hoboken (2008)
4. Kobetic, R., To, C., Schnellenberger, J., Audu, M., Bulea, T., Gaudio, R., Pinault, G., Tashman, S., Triolo, R.J.: Development of hybrid orthosis for standing, walking, and stair climbing after spinal cord injury. J. Rehabil. Res. Dev. **46**(3), 447–462 (2009)
5. Tucker, M.R., Olivier, J., Pagel, A., Bleuer, H., Bouri, M., Lambercy, O., Milln, J.R., Riener, R., Vallery, H., Gassert, R.: Control strategies for active lower extremity prosthetics and orthotics: a review. J. Neuroeng. Rehabil. **12**(1) (2015)
6. Novak, D., Riener, R.: A survey of sensor fusion methods in wearable robotics. Robot. Auton. Syst. **73**, 155–170 (2015)
7. Huang, H., Zhang, F., Hargrove, L.J., Dou, Z., Rogers, D.R., Englehart, K.B.: Continuous locomotion-mode identification for prosthetic legs based on neuromuscular-mechanical fusion. IEEE Trans. Biomed Eng. **58**(10), 2867–2875 (2011)
8. Gorsic, M., Kamnik, R., Ambrozic, L., Vitiello, N., Lefeber, D., Pasquinia, G., Munih, M.: Online phase detection using wearable sensors for walking with a robotic prosthesis. Sensors **14**(2), 2776–2794 (2014)
9. Ambrozic, L., Gorsic, M., Geeroms, J., Flynn, L., Molino, Lova R., Kamnkik, R., Munih, M., Vitiello, N.: CYBERLEGs: a user-oriented robotic transfemoral prosthesis with whole-body awareness control. IEEE Robot. Autom. Mag. **21**(4), 82–93 (2014)
10. Yuan, K., Wang, Q., Wang, L.: Fuzzy-logic-based terrain identification with multisensor fusion for transtibial amputees. IEEE Trans. Mechatron. **20**(2), 618–630 (2015)

11. Chen, B., Wang, X., Huang, Y., Wei, K., Wang, Q.: A foot-wearable interface for locomotion mode recognition based on discrete contact force distribution. Mechatronics **32**, 12–21 (2015)
12. Sup, F., Varol, H.A., Goldfarb, M.: Upslope walking with a powered knee and ankle prosthesis: initial results with an amputee subject. IEEE Trans. Neural Syst. Rehabil. **19**(1), 71–78 (2011)
13. Tkach, D.C., Hargrove, L.J.: Neuromechanical sensor fusion yields highest accuracies in predicting ambulation mode transitions for transtibial amputees. In: Proceedings of the 35th Annual International Conference of the IEEE Engineering in Medicine and Biology Society, pp. 3074–3077 (2013)
14. Young, A.J., Simon, A., Hargrove, L.J.: An intent recognition strategy for transfemoral amputee ambulation across different locomotion modes. In: Proceedings of the 35th Annual International Conference of the IEEE Engineering in Medicine and Biology Society, pp. 1587–1590 (2013)
15. Giovacchini, F., Vannetti, F., Fantozzi, M., Cempini, M., Cortese, M., Parri, A., Yan, T., Lefeber, D., Vitiello, N.: A light-weight active orthosis for hip movement assistance. Robot. Auton. Syst. **73**, 123–134 (2015)
16. Pratt, G.A., Williamson, M.M.: Series elastic actuators. In: Proceedings of the IEEE/RSJ International Conference on Intelligent Robots and Systems, pp. 399–406 (1995)
17. Cortes, C., Vapnik, V.: Support-vector network. Mach. Learn. **20**(3), 273–297 (1995)
18. Bevilacqua, V., Pannarale, P., Abbrescia, M., Cava, C., Paradiso, A., Tommasi, S.: Comparison of data-merging methods with SVM attribute selection and classification in breast cancer gene expression. BMC Bioinform. **13**(7) (2012)
19. Chen, P., Liu, S.: An improved DAG-SVM for multi-class classification. In: Proceedings of the 5th International Conference on Natural computation, pp. 460–462 (2009)
20. Geisser, S.: Predictive Inference. Chapman & Hall, London (1993)
21. Parri, A., Yuan, K., Marconi, D., Yan, T., Munih, M., Molino Lova, R., Vitiello, N., Wang, Q.: Real-time hybrid ecological intention decoding for lower-limb wearable robots. IEEE Trans. Mechatron. (Accepted for publication)
22. Jang, J., Kim, K., Lee, J., Lim, B., Shim, Y.: Online gait task recognition algorithm for hip exoskeleton. In: Proceedings of the IEEE/RSJ International Conference on Intelligent Robots and Systems, pp. 5327–5332 (2015)

Object Segmentation in Depth Maps with One User Click and a Synthetically Trained Fully Convolutional Network

Matthieu Grard, Romain Brégier, Florian Sella,
Emmanuel Dellandréa and Liming Chen

Abstract With more and more household objects built on planned obsolescence and consumed by a fast-growing population, hazardous waste recycling has become a critical challenge. Given the large variability of household waste, current recycling platforms mostly rely on human operators to analyze the scene, typically composed of many object instances piled up in bulk. Helping them by robotizing the unitary extraction is a key challenge to speed up this tedious process. Whereas supervised deep learning has proven very efficient for such object-level scene understanding, e.g., generic object detection and segmentation in everyday scenes, it however requires large sets of per-pixel labeled images, that are hardly available for numerous application contexts, including industrial robotics. We thus propose a step towards a practical interactive application for generating an object-oriented robotic grasp, requiring as inputs only one depth map of the scene and one user click on the next object to extract. More precisely, we address in this paper the middle issue of object segmentation in top views of piles of bulk objects given a pixel location, namely seed, provided interactively by a human operator. We propose a two-fold framework for generating edge-driven instance segments. First, we repurpose a state-of-the-art fully convolutional object contour detector for seed-based instance segmentation by introducing the notion of *edge-mask duality* with a novel patch-free and contour-oriented loss function. Second, we train one model using only synthetic scenes, instead of manually labeled training data. Our experimental results show that considering edge-mask duality for training an encoder-decoder network, as we suggest, outperforms a state-of-the-art patch-based network in the present application context.

M. Grard (✉) · R. Brégier · F. Sella
Siléane, 17 Rue Descartes, 42000 St Étienne, France
e-mail: m.grard@sileane.com

M. Grard · E. Dellandréa · L. Chen
Université de Lyon, CNRS, École Centrale de Lyon, LIRIS UMR5205,
69134 Lyon, France

R. Brégier
Université Grenoble Alpes, Inria, CNRS, Grenoble INP, LIG, 38000 Grenoble, France

© Springer International Publishing AG, part of Springer Nature 2019
F. Ficuciello et al. (eds.), *Human Friendly Robotics*, Springer Proceedings
in Advanced Robotics 7, https://doi.org/10.1007/978-3-319-89327-3_16

207

Keywords Interactive instance segmentation · Supervised pixel-wise learning
Synthetic training images

1 Introduction

Waste recycling is one of the main challenges lying ahead, as the ever-growing amount of obsolete household objects and industrial materials outstrips the deployed human resources. Current state legislations and policies thus entice industrialists to develop robotized waste sorting, which stands out as a competitive approach to protect humans from hazardous waste collection sites (toxic emissions, radioactive radiations, etc.). However, given the large variability of elements to sort, and the cluttered nature of waste, human expertise is hardly dispensable despite the great advances in machine learning. Towards a fast and accurate processing, enhancing human operators decisions with computer vision-guided robotic grasping seems therefore one of the current most suitable strategies. Numerous works on supervised learning for object-wise scene understanding in images of everyday scenes have indeed demonstrated remarkable results for real-time applications: *generic object detection*, i.e., generating category-free object proposals, either as bounding box proposals [12, 24], or as binary masks, namely segment proposals [10, 14, 22, 23]; *semantic segmentation*, i.e., assigning an instance-agnostic object category to each pixel [16, 18]; *object contour detection*, i.e., assigning to each pixel a probability of being an object boundary [5, 30].

In this work, we aim at extracting one by one each instance of a dense pile of objects in random poses. As illustrated by Fig. 1, we consider the following scenario: (1) a depth-augmented image of the scene is captured; (2) a human operator clicks on the next object to extract; (3) the selected object is automatically delineated; (4) robotic grasps on the segmented object are automatically detected. This paper describes the third step, i.e., the segmentation of the selected object. As the ultimate goal is to successfully grasp each item of the scene, object instances must be correctly distinguished despite potential occlusion and entanglement, which is hardly achievable without an explicit notion of object. In the context of waste sorting, prior knowledge is not available: the target objects to be extracted, as well as their corresponding model, are unknown. Differently from industrial bin-picking of many instances of a known manufactured object, traditional pose estimation techniques [4] cannot be applied. We thus adapt object segmentation methods used for everyday image analysis to the context of industrial robotic picking. Indeed, state-of-the-art techniques [10, 22] first perform a coarse localization by extracting a set of image patches assumed to contain one instance. While this approach is relevant for the typical foreground/background paradigm inherent to most indoor and outdoor scenes, it can be ambiguous when the scene is only composed of objects densely stacked on top of each other, as a patch may be equally shared by many objects (cf. Fig. 2). The proposed approach alleviates this problem by considering a pixel included in the object, instead of a bounding box. Interestingly, one can notice in the present

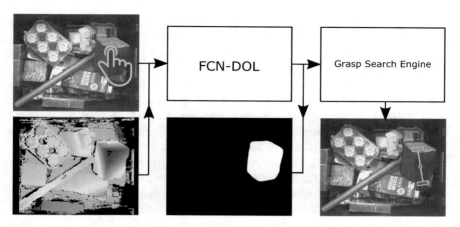

Fig. 1 A test application of our approach on a real robotic setup: (1) A user clicks on the object to extract. (2) The proposed model (FCN-DOL) delineates the selected object in a single forward pass. (3) An off-the-shelf grasp search engine can thus generate a relevant object-oriented robotic grasp

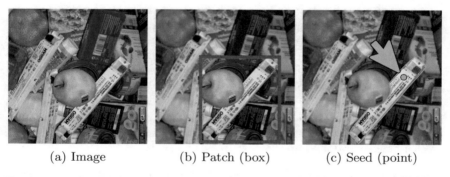

(a) Image   (b) Patch (box)   (c) Seed (point)

Fig. 2 A use case showing that it is less ambiguous to consider the task "Segment the object including this pixel" (**c**) rather than "Segment the object inside this box" (**b**) as a patch may be equally shared by several objects

application context that inter-instance boundaries are complementary to instance segments, since the background is generally fully hidden and all objects are targets. This suggests besides that object contour detection may be used as a backbone for object segmentation. In the following, we refer to this property as *edge-mask duality*.

Moreover, state-of-the-art learning-based edge detection and segmentation tasks require training datasets annotated on a per-pixel basis. Despite many efforts [7], notably for real indoor scenes featuring household objects [15, 29], and table-top sparse clutters [25], none matches the camera point of view, neither the typical scene setup of encountered waste sorting applications. Although collecting adequate data is certainly not a problem given the recent spread of affordable RGB-D sensors, annotating reliably new images remains tedious, as current methods consist in time-consuming manual labeling. Real image datasets are therefore limited in terms of content, and costly to extend. This makes the availability of appropriate training data a factor hindering the widespread deployement of pixel-wise learning-based techniques in industrial environments.

In this paper, towards an interactive application for robotized waste sorting, we address the middle issue of learning object segmentation in depth maps, given one user click per object to segment. We aim at demonstrating that leveraging edge-mask duality with synthetic depth maps for end-to-end instance segmentation enables higher boundary precision than a patch-based approach, while ensuring a framework fully pluggable in an interactice interface. Starting from the observations that (1) realistic RGB-D images of synthetic scenes can be automatically generated by simulation, (2) learning from synthetic depth-augmented images has already shown promising results for semantic segmentation of indoor scenes [8] and disparity estimation [19], (3) fully convolutional networks trained end to end for object contour detection has proven very effective [30], our contribution is two-fold:

- **A synthetic expandable dataset for industrial robotic picking**. With custom scripts on top of Blender [2], we produce realistic rendered RGB-D images of object instances stacked in bulk, thus ensuring time-saving and error-free annotations used as pixel-wise ground truth during training.
- **A scheme to train a fully convolutional edge detector for seed-based instance segmentation**. First, we show that object contour detectors can be trained on synthetic depth maps while performing almost equivalently on real data at inference time. Second, we introduce edge-mask duality with a novel dual-objective loss function to train a fully convolutional encoder-decoder in segmenting objects clicked interactively by a human operator. We show that the proposed approach, more suited for a practical human-robot interface, outperforms a state-of-the-art patch-based network when objects are piled up in bulk.

Our paper is organized as follows. After reviewing the state of the art in Sect. 2, we describe the proposed training scheme in Sect. 3 and our training data production pipeline in Sect. 4. Then, Sect. 5 details our experimental protocol. Results are discussed in Sect. 6.

2 Related Work

2.1 Synthetic Training Data

To our knowledge, learning from synthetic RGB-D images for pixel-wise object segmentation tasks has yet received little attention. In the context of generic object detection and recognition, a first kind of approach consists in generating synthetic images by projecting 3D models onto natural "background" images [20, 28]. However, these synthetic images contain by construction at most a few object instances that are implicitly assumed to be distinguishable from their environment, thus excluding many situations like the present case of dense stack of object instances. A second type of approach consists in simulating scenes from scratch. Mainly dedicated to semantic segmentation, synthetic datasets of annotated urban [27] and indoor [9] scenes have recently emerged. Following the creation of these datasets, it has been shown that depth-based semantic segmentation of indoor scenes with a convolutional neural network (CNN) can be boosted by using synthetic images as training data and ten times less real images only for fine-tuning [8]. Also, in the footsteps of FlowNet [6] for optical flow estimation using synthetic images, training a CNN only on sufficiently realistic rendered image patches has again recently proved efficient for the purposes of disparity estimation on outdoor scenes with DispNet [19]. In continuity with the latter simulation-based approaches, we explore the context of robotic picking when the target objects are undefined. We show that real depth maps, whose costly pixel-wise ground-truth annotations are unsustainable in industrial environments, may be fully replaced by rendered depth maps when feeding structured random forests or fully convolutional networks for object contour detection during training.

2.2 Instance Segmentation

Due to a lack of large real-world RGB-D datasets, instance segmentation algorithms have been mostly performed on RGB-only images of everyday scenes, augmented with crowd-sourced pixel-wise annotations [17]. Concurrently with the emergence of end-to-end deep learning for instance segmentation, images were typically represented as a graph of superpixels, often built from boundary-related cues. To this end, popular edge detectors, using structured random forests to classify patches [31] or lately end-to-end fully convolutional "encoder-decoder" networks [30], were employed to give edge-preserving starting cues. Unsupervised instance proposals generation then typically consisted in multi-scale combinatorial grouping [23] based upon generic hand-crafted features, e.g., instance candidate's area or perimeter. Later, learning to propose instance candidates was notably introduced with conditional random fields (CRFs) [14], that are jointly trained for binary graph cuts while maximizing global recall. Lately, Facebook AI Research released a network that predicts scored instance proposals from image patches [21], trained end to end on the MS COCO dataset [17], and outperforming previous state-of-the-art approaches. A top-down refinement scheme was further proposed [22] to refine the boundaries of the predicted masks, since without explicit notion of object contour during training, feed-forward deep architectures generally fail to encode accurate boundaries due to progressive down-sampling. Within the context of semantic instance segmentation, recurrent neural networks have been employed to sequentially segment each instance belonging to one inferred category [26], by heading fully convolutional networks for semantic segmentation with long short-term memory (LSTM) layers, so as to enable a recurrent spatial inhibition. In contrast to previous approaches, we leverage the duality between inter-instance boundaries and instance masks (described in Sect. 3) directly within a fast feed-forward architecture by introducing an explicit object contour learning during training for instance segmentation, so as to obtain masks with sharp and consistent boundaries in the case of objects piled up in bulk.

3 Leveraging Edge-Mask Duality

Motivated by the many recent works on feed-forward fully convolutional architectures that demonstrated strong learning capabilities and fast inference time to be part of a responsive interactive interface, we follow the idea that when the scene is full of target objects, typically in waste sorting applications, the inter-instance boundaries and the set of instance masks are dual representations, what we call *edge-mask duality*. This interestingly suggests that feature encoding-decoding for object contour detection may be used as backbone for instance segmentation. One may indeed expect a better object pose invariance and a finer segmentation when learning local inter-instance boundaries instead of the spatial continuity inside the masks. We thus propose to repurpose a state-of-the-art encoder-decoder object contour detector to predict instance segments by appending during training to the input and the output respectively a seed pixel and the instance mask including the seed. State-of-the-art approaches [10, 22] link a patch, either randomly generated or predicted, to one instance, which requires size priors and may be ambiguous (cf. Fig. 2). We use instead a translation centering a pixel location assumed to be included in one instance, namely seed, here provided by an external user, as illustrated by Fig. 3.

Formally, suppose we have a set of annotated depth maps. For one depth map, let $\mathcal{G}$ be the ground-truth instance label map and $\mathcal{E}$ the ground-truth inter-instance contour map, that is, for each pixel $\mathbf{p}$,

$$\mathcal{E}_\mathbf{p} = \begin{cases} 1 & \text{if } \exists (\mathbf{q}, \mathbf{q}') \in \mathcal{N}_\mathbf{p}^2 \text{ such that } \mathcal{G}_\mathbf{q} \neq \mathcal{G}_{\mathbf{q}'} \\ 0 & \text{otherwise} \end{cases} \tag{1}$$

where $\mathcal{N}_\mathbf{p}$ denotes the 8-connected neighborhood of pixel $\mathbf{p}$. In other words, $\mathcal{E}_\mathbf{p}$ is the expected probability that $\mathbf{p}$ is an instance boundary. Let $\{\mathcal{M}_k\}_k$ be the set of the corresponding ground-truth instance binary masks. One can simply define the instance mask $\mathcal{M}_\mathbf{s}$ at seed pixel $\mathbf{s}$ as a two-operation process on the contour map:

$$\mathcal{M}_\mathbf{s} = \mathcal{C}(\bar{\mathcal{E}}, \mathbf{s}) \tag{2}$$

where $\mathcal{C}(\mathcal{X}, \mathbf{s})$ is the connected subset of $\mathcal{X}$ including $\mathbf{s}$, and $\bar{\mathcal{X}}$ the complementary of $\mathcal{X}$. $\mathcal{M}_\mathbf{s}$ thus defines the expected probability for each pixel $\mathbf{p}$ to belong to the instance including the seed $\mathbf{s}$. Reciprocally, the instance contour map can be trivially derived from the set of instance masks.

Feed-forward networks require one-to-one or many-to-one relationships between the training inputs and expected outputs, however in the context of instance segmentation, the corresponding set of instances is of varying cardinality. Circumventing this issue thus requires each expected instance mask to be associated with one different input. State-of-the-art approaches [22] implement one-to-one relationships by defining a set of patches, assuming that one patch contains one instance. In this work, we instead consider a set $\mathcal{S}$ of seed pixels and propose the following mapping:

$$\begin{aligned} \mathcal{S} &\to \quad \{\mathcal{M}_k\}_k \\ \mathbf{s} &\mapsto \quad \mathcal{M}_{\mathbf{s}} \end{aligned} \tag{3}$$

so that each seed pixel maps to the instance it belongs to. Exploiting the translational invariance of convolutional networks and Eq. 3, we can then define a pixel-wise training set to use encoder-decoder networks for instance segmentation, by mapping one depth map and one seed $\mathbf{s}$ to one contour map $\mathcal{E}$ and one instance mask $\mathcal{M}_{\mathbf{s}}$, as depicted by Fig. 3. At training time, we select randomly a few seeds inside each ground-truth instance. At inference time, the user provides a seed by clicking on the instance.

To introduce the edge-mask duality in the learning, we train the encoder-decoder network end to end by minimizing the following sum of pixel-wise logistic loss functions

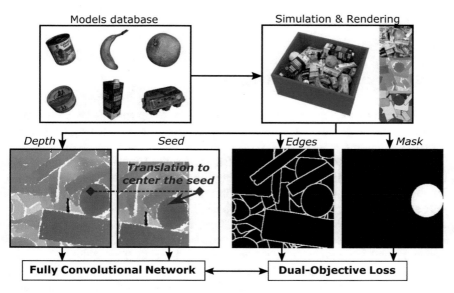

Fig. 3 Our training scheme to generate instance segments by introducing the notion of *edge-mask duality* with a novel loss function. A fully convolutional encoder-decoder network is fed with one image and its translation centering the user click on one instance, namely seed, to predict the overall inter-instance boundary map and the instance mask including the input seed. The proposed dual-objective loss drives the network to explicitly learn segmentation from boundaries

$$\mathcal{L}(\theta) = \frac{1}{N} \sum_{n=1}^{N} \left\{ \ell(\lambda_e, \mathcal{E}^{(n)}, \hat{\mathcal{E}}^{(n)}) + \sum_{s \in \mathcal{S}} \ell(\lambda_m, \mathcal{M}_{\mathbf{s}}^{(n)}, \hat{\mathcal{M}}_{\mathbf{s}}^{(n)}) \right\} \tag{4}$$

where θ denotes the network's parameters, N the number of training samples, ℓ has the following form:

$$\ell(\lambda, \mathcal{Y}, \hat{\mathcal{Y}}) = -\sum_{\mathbf{p}} \left\{ (1 - \mathcal{Y}_{\mathbf{p}}) \log(1 - \sigma(\hat{\mathcal{Y}}_{\mathbf{p}})) \right.$$
$$\left. + \lambda \ \mathcal{Y}_{\mathbf{p}} \log(\sigma(\hat{\mathcal{Y}}_{\mathbf{p}})) \right\} \tag{5}$$

$\mathcal{Y}_{\mathbf{p}} \in \{0, 1\}$ and $\hat{\mathcal{Y}}_{\mathbf{p}} \in \mathbb{R}$ are respectively the expected output and the encoder-decoder network's response at pixel $\mathbf{p}$, λ is a trade-off parameter between "object contour" or "instance mask" pixels and inactive ones, and σ is the sigmoid function. While the term $\ell(\lambda_e, \mathcal{E}, \hat{\mathcal{E}})$ focuses the network's attention on boundaries, the term $\ell(\lambda_m, \mathcal{M}_{\mathbf{s}}, \hat{\mathcal{M}}_{\mathbf{s}})$ enables to learn per-seed connectivity, so that encoding-decoding for predicting instance contours on the whole image and instance masks is jointly learned, as suggested by Eq. 2. In practice, we set $\lambda_e = 10$ and $\lambda_m = 1$. By definition, $\mathcal{L}$ is fully differentiable as a linear combination of logistic losses. With the proposed loss function, instance segmentation is performed in one forward pass while learning explicitly driven by inter-instance boundaries.

4   Synthesizing Training Images

To train the network, we generate a synthetic dataset using Blender [2] with custom code to simulate scenes of objects in bulk and render the corresponding RGB-D top views.

Simulation We model a static bin, an RGB-D sensor and some objects successively dropped above the bin in random pose using Blender's physics engine. We add some intraclass geometric variability if needed, by applying isotropic and anisotropic random scaling. In both our real robotic setup and Blender, depth is recovered using an active binocular stereoscopic system composed of two RGB cameras and a pseudo-random pattern projector to add artificial texture to the scene. Calibration, rectification and matching are performed using standard off-the-shelf algorithms.

Rendering Once all instances have been dropped, we render each camera view using Cycles render engine and process the pair of rendered images in the exact same way as a pair of real images. The depth sensor's noise model is thus implicitly embedded during stereo matching, which gives realistic depth maps without additional post-processing, contrary to [8]. Figure 4 shows a qualitative comparison between real and synthetic images from our dataset.

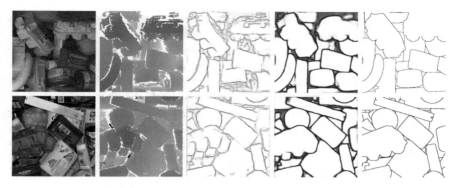

Fig. 4 Detected object contours on real (first row) and synthetic (second row) test images, both using the same model trained only on our synthetic depth maps. From left to right: RGB, depth, contours using a structured random forest (RF) [5], contours using a fully convolutional network (FCN) [30], ground truth

5 Experimental Setup

We conduct two sets of experiments: (1) we train two state-of-the-art object contour detectors—a structured random forest (RF) [5] and a fully convolutional network (FCN) [30]—on synthetic depth maps to achieve equivalent performance on real data at inference time; (2) we show that learning instance segmentation by training a fully convolutional encoder-decoder network with our novel dual-objective loss function (FCN-DOL) outperforms a patch-based network (DeepMask) [22] in terms of boundary precision, while requiring only seed pixels instead of patches. In all our experiments, we focus on synthetic depth as rendered RGB is less realistic and depth conveys more generalizable information.

Data preparation To evaluate our synthetic training, we capture 25 RGB-D top views of real-world scenes featuring 15 household objects. Four expert human annotators manually label each image with the ground-truth contours. Using an off-the-shelf 3D scanner, we scan each of the 15 objects, and generate 1,000 synthetic images of multi-object scenes, i.e., scenes with many instances of many objects, and 1,500 mono-object scenes (100 per object), i.e., scenes with many instances of one object. The multi-object scenes are randomly divided into 3 subsets with the following distribution: 400 for training, 200 for validation, and 400 for test. All mono-object scenes are used only for test. In addition, 5 synthetic and 5 real images are manually labeled separately, each one by 3 different annotators, to evaluate the quality of manual annotation and the impact of ground truth on contour detection performances. We limit the manual annotation to a few images due to the burden of the task, that required 68 h in total.

Performance metrics For contour detection, we use standard metrics from [1]: the best F-score, and the corresponding recall and precision, with a fixed threshold on dataset scale. We add the *synthetic gap*, i.e. the absolute difference between scores on real (R) and synthetic (S) test images. For instance segmentation, we compute two quantities: the average intersection over union ratio (IoU) over the best ground-truth-matching proposal set; and the boundary precision on instance scale using the same metric as for contour detection. Given one image, a proposal best matches a ground-truth instance if it has the highest IoU among all proposals for one of the ground-truth instances, and if the corresponding IoU is higher than 50%.

Settings for evaluating the use of synthetic images We train two object contour detectors, state-of-the-art in their kind: a structured random forest (RF) from [5] of 4 trees and a fully convolutional network (FCN) from [30] with a VGG-16-based encoder-decoder architecture—on synthetic depth maps to perform equivalently on real data at inference time. For RF, the trees are trained on 10^6 patches extracted from 100 synthetic multi-object images (10^4 patches—as many positive as negative ones—per training image). Multi-scale option and non-maximal suppression are not enabled at test time. Other parameters are set to their default value. For FCN, we perform 180 epochs on 400 synthetic images, using Caffe [13]. Depth is introduced as a new input layer replacing the BGR inputs, therefore both encoder and decoder's weights are learned. Weights are initialized like in [30], except that two channels of the input layer are removed. We set a constant learning rate of 10^{-4} and a weight decay of 10^{-4}. For both RF and FCN, tests are performed on the 25 manually annotated real depth maps (R-tests), and on 25 synthetic depth maps (S-tests) of multi-object scenes. Human annotators are evaluated on 5 real and 5 synthetic images.

Settings for evaluating the use of edge-mask duality We train two feed-forward deep architectures only on synthetic depth maps of multi-object scenes: (i) the state-of-the-art patch-based network referred to as DeepMask [22], built on a 50-layer residual encoder [11] and taking patches as input; (ii) the previous encoder-decoder network referred to as FCN [30] but trained using our dual-objective loss function (FCN-DOL). For these trainings, we removed all partially occluded instances from ground truth, so that each model is trained only on instances whose mask is connected and of reasonably large area, i.e., graspable instances. DeepMask is initialized with the weights pretrained on MS COCO, and fine-tuned by performing 100 epochs on our 400 synthetic training images, with a constant learning rate of 10^{-4} and data augmentation (shift and scale jittering). FCN-DOL is initialized with our FCN model for contour detection. We add one input layer, corresponding to the translated image that centers the seed pixel, and one output layer, corresponding the object mask including the seed. We then perform 120 epochs driven by our dual-objective loss function, with a constant learning rate of 10^{-4}, using 2 random pixel seeds per ground-truth instance and data augmentation (4 rotations and horizontal mirror). At test time, we define a constant regular grid of seeds over the whole image. For each image, we obtain a set of object mask proposals by forwarding each seed of the grid, and binarizing the continuous output mask with a threshold of 0.8. For both DeepMask and FCN-DOL, tests are performed on 400 synthetic depth maps of multi-object scenes and 1,500 synthetic images of mono-object scenes.

6 Results

Synthetic over real training images Using simulation enables significant time-savings: annotating a real image takes 40 min in average while generating a synthetic view about 5 min (with a quad-core 3.5 GHz Intel Xeon E5 for physics simulation, and a Nvidia Quadro M4000 for GPU rendering), thus 8 times faster. Interestingly, both edge detectors trained on synthetic depth maps still achieve good results in practice when tested on our real data, producing contour maps visually similar to those obtained with a synthetic input, as illustrated in Fig. 4. However, Table 1 shows synthetic gaps of 16 and 18 points in F-score inducing that our synthetic data is not realistic enough, here in terms of depth noise modeling. Nevertheless, this gap can be partially relativised, considering the non-zero gap (12 points) achieved by humans. S-tests are indeed conducted against perfect ground truth while R-tests against error-prone annotations, as humans miss hardly detectable edges in noisy areas. Bridging the gap with more realism should be part of a future work to limit any overtraining in the synthetic domain.

Seed-based edge-driven over patch-based instance segments As shown by Fig. 5, our dual-objective loss for instance segmentation by learning explicitly inter-object boundaries leads to sharper mask boundaries than using a state-of-the-art patch-based network, when the scene is a pile of bulk objects as it is the case in a waste sorting application. These results are corroborated by Table 2, with a gain of 39 points in boundary precision. In addition, our model performs equally on mono-object setups though it was trained only on multi-object scenes, whereas the patch-based network's performance drops by 18 points. This suggests that introducing the edge-mask duality at training time enables a better generalization on new arrangements. In a real interactive interface, the human-guided network helps to relevantly reduce the grasp search space as illustrated by Fig. 1. Although the proposed model implies one forward pass per seed, this is not problematic in our application context as we trust the user to perform a relevant click. Our experimental dataset should nevertheless be extended to a much larger database of 3D models such as ShapeNet [3] to investigate

Table 1 Comparative results for instance contour detection on real (R) and synthetic (S) images, using a random forest (RF) [5] and a fully convolutional network (FCN) [30] both trained only on our synthetic depth maps. The synthetic gap (Gap) is the absolute difference between R and S-scores. Humans are given RGB since they are unable to distinguish instances only from depth

	Modality	Train	Test								
			Recall			Precision			F-score		
			25R	25S	Gap	25R	25S	Gap	25R	25S	Gap
RF [5]	Depth	100S	0.54	0.70	**0.16**	0.53	0.69	**0.16**	0.53	0.69	**0.16**
FCN [30]	Depth	400S	0.43	0.60	**0.17**	0.78	0.93	**0.15**	0.55	0.73	**0.18**
Humans	RGB	–	0.48	0.65	**0.17**	0.98	0.92	**0.06**	0.64	0.76	**0.12**

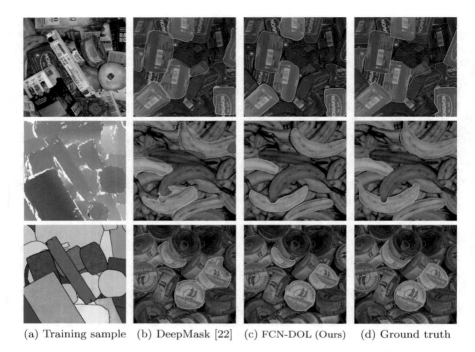

(a) Training sample   (b) DeepMask [22]   (c) FCN-DOL (Ours)   (d) Ground truth

Fig. 5 Best matching instance proposals on synthetic test depth maps. With one depth-only training on multi-object scenes (**a**), our edge-driven model (**c**) gives more edge-preserving instances and better generalizes to unknown arrangements, in contrast to patch-based networks (**b**)

Table 2 Intersection over Union (IoU) and boundary precision of the best matching instances on synthetic multi-object (Mu) and mono-object (Mo) scenes

	Train	400Mu		1500Mo	
		Average best IoU	Boundary precision	Average best IoU	Boundary precision
DeepMask [22]	400Mu	0.83	0.51	0.88	0.33
FCN-DOL (Ours)	400Mu	**0.89**	**0.90**	**0.88**	**0.90**

the generalization on new objects, as our current images lack variability in terms of object category.

Limitations of the proposed model and future work The proposed model relies on a state-of-the-art encoder-decoder structure with 130M parameters against less than 60M parameters for DeepMask. This, and the lack of residual blocks, hinder the training, which explains why we performed a two-phase training with three times more fine-tuning epochs than DeepMask in total to reach a stable accuracy. Building a new encoder-decoder architecture with residual blocks and less filters for the layer connecting the encoder and the decoder—currently 79% of the parameters to train—could alleviate this issue. Also, thanks to the translation invariance of fully

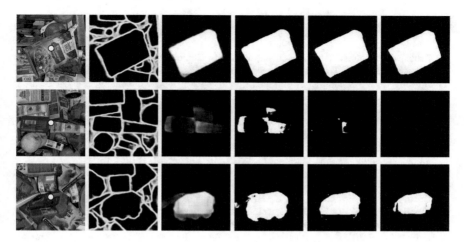

Fig. 6 Examples of clicks (the white disks) using the proposed model. From left to right: RGB, predicted edges, predicted continuous mask, binarized masks with respectively a threshold of 0.2, 0.5, 0.8. From top to bottom: a click near the instance's centroid; a click on a boundary; a click near a boundary

convolutional networks, any pixel belonging to the object may be selected, and in practice, we can expect a reasonable seed, as a human user instinctively clicks near the object's centroid. As depicted in Fig. 6, a click on a boundary produces a low-contrasted output as the selected seed doesn't belong to any instance. A click near a boundary returns a continuous segmentation map more sensitive to the binarization threshold. The robustness on this point could be increased with a smarter seed selection at training time.

7 Conclusion

We presented a simulation-based framework towards an interactive application for robotized waste sorting. Realistic synthetic RGB-D images are generated to train an encoder-decoder network for edge-driven instance segmentation in depth maps. Assuming edge-mask duality in the images, we introduce a novel dual-objective loss function to explicitly focus the learning on inter-instance boundaries. Given seeds provided interactively by a human operator, the network thus produces instance masks with sharper boundaries than a state-of-the-art patch-based approach, in a single forward pass, and better generalizes to unknown object arrangements. Its deep fully convolutional architecture enables strong learning capabilities into a responsive human-robot interface to speed up the overall process of selecting, extracting and sorting objects.

References

1. Arbeláez, P., Maire, M., Fowlkes, C., Malik, J.: Contour detection and hierarchical image segmentation. IEEE Trans. Pattern Anal. Mach. Intell. (TPAMI) **33**(5), 898–916 (2011)
2. Blender—A 3D Modelling and Rendering Package. Blender Foundation, Blender Institute, Amsterdam (2016)
3. Chang, A.X., Funkhouser, T.A., Guibas, L.J., Hanrahan, P., Huang, Q.-X., Li, Z., Savarese, S., Savva, M., Song, S., Su, H., Xiao, J., Yi, L., Yu, F.: ShapeNet: an information-rich 3D model repository (2015). CoRR arXiv:abs/1512.03012
4. Choi, C., Taguchi, Y., Tuzel, O., Liu, M.-Y., Ramalingam, S.: Voting-based pose estimation for robotic assembly using a 3D sensor. In: ICRA, pp. 1724–1731. IEEE (2012)
5. Dollár, P., Zitnick, C.L.: Fast edge detection using structured forests (2014). CoRR arXiv:abs/1406.5549
6. Dosovitskiy, A., Fischer, P., Ilg, E., Husser, P., Hazirbas, C., Golkov, V., van der Smagt, P., Cremers, D., Brox, T.: FlowNet: learning optical flow with convolutional networks. In: ICCV, pp. 2758–2766. IEEE Computer Society (2015)
7. Firman, M.: RGBD datasets: past, present and future. In: The IEEE Conference on Computer Vision and Pattern Recognition (CVPR) Workshops, June 2016
8. Handa, A., Patraucean, V., Badrinarayanan, V., Stent, S., Cipolla, R.: Understanding realworld indoor scenes with synthetic data. In: CVPR, pp. 4077–4085. IEEE Computer Society (2016)
9. Handa, A., Patraucean, V., Stent, S., Cipolla, R.: SceneNet: an annotated model generator for indoor scene understanding. In: Kragic, D., Bicchi, A., Luca, A.D. (eds.) ICRA, pp. 5737–5743. IEEE (2016)
10. He, K., Gkioxari, G., Dollár, P., Girshick, R.: Mask-RCNN. In: ICCV (2017)
11. He, K., Zhang, X., Ren, S., Sun, J.: Deep residual learning for image recognition. In: CVPR, pp. 770–778. IEEE Computer Society (2016)
12. Hosang, J.H., Benenson, R., Dollr, P., Schiele, B.: What makes for effective detection proposals? (2015). CoRR arXiv:abs/1502.05082
13. Jia, Y., Shelhamer, E., Donahue, J., Karayev, S., Long, J., Girshick, R., Guadarrama, S., Darrell, T.: Caffe: convolutional architecture for fast feature embedding (2014). arXiv:1408.5093
14. Krahenbuhl, P., Koltun, V.: Learning to propose objects. In: CVPR, pp. 1574–1582. IEEE (2015)
15. Lai, K., Bo, L., Ren, X., Fox, D.: A large-scale hierarchical multi-view RGB-D object dataset. In: ICRA, pp. 1817–1824. IEEE (2011)
16. Lin, G., Shen, C., van den Hengel, A., Reid, I.: Efficient piecewise training of deep structured models for semantic segmentation. In: CVPR, pp. 3194–3203. IEEE Computer Society (2016)
17. Lin, T.-Y., Maire, M., Belongie, S., Hays, J., Perona, P., Ramanan, D., Dollár, P., Zitnick, C.L.: Microsoft COCO: common objects in context. In: Fleet, D.J., Pajdla, T., Schiele, B., Tuytelaars, T. (eds.) ECCV (5). Lecture Notes in Computer Science, vol. 8693, pp. 740–755. Springer (2014)
18. Long, J., Shelhamer, E., Darrell, T.: Fully convolutional networks for semantic segmentation. In: CVPR, pp. 3431–3440. IEEE (2015)
19. Mayer, N., Ilg, E., Husser, P., Fischer, P., Cremers, D., Dosovitskiy, A., Brox, T.: A large dataset to train convolutional networks for disparity, optical flow, and scene flow estimation. In: CVPR, pp. 4040–4048. IEEE Computer Society (2016)
20. Peng, X., Sun, B., Ali, K., Saenko, K.: Learning deep object detectors from 3D models. In: ICCV, pp. 1278–1286. IEEE Computer Society (2015)
21. Pinheiro, P.H.O., Collobert, R., Dollár, P.: Learning to segment object candidates. In: Cortes, C., Lawrence, N.D., Lee, D.D., Sugiyama, M., Garnett, R. (eds.) NIPS, pp. 1990–1998 (2015)
22. Pinheiro, P.O., Lin, T.-Y., Collobert, R., Dollár, P.: Learning to refine object segments. In: Leibe, B., Matas, J., Sebe, N., Welling, M. (eds.) ECCV (1). Lecture Notes in Computer Science, vol. 9905, pp. 75–91. Springer (2016)

23. Pont-Tuset, J., Arbelaez, P., Barron, J.T., Marqus, F., Malik, J.: Multiscale combinatorial grouping for image segmentation and object proposal generation (2015). CoRR arXiv:abs/1503.00848
24. Ren, S., He, K., Girshick, R., Sun, J.: Faster R-CNN: towards real-time object detection with region proposal networks. In: Cortes, C., Lawrence, N.D., Lee, D.D., Sugiyama, M., Garnett, R. (eds.) NIPS, pp. 91–99 (2015)
25. Richtsfeld, A., Morwald, T., Prankl, J., Zillich, M., Vincze, M.: Segmentation of unknown objects in Indoor environments. In: IROS, pp. 4791–4796. IEEE (2012)
26. Romera-Paredes, B., Torr, P.H.S.: Recurrent instance segmentation. In: Leibe, B., Matas, J., Sebe, N., Welling, M. (eds.) ECCV (6). Lecture Notes in Computer Science, vol. 9910, pp. 312–329. Springer (2016)
27. Ros, G., Sellart, L., Materzynska, J., Vzquez, D., Lopez, A.M.: The SYNTHIA dataset: a large collection of synthetic images for semantic segmentation of urban scenes. In: CVPR, pp. 3234–3243. IEEE Computer Society (2016)
28. Rozantsev, A., Lepetit, V., Fua, P.: On rendering synthetic images for training an object detector. Comput. Vis. Image Underst. **137**, 24–37 (2015)
29. Silberman, N., Hoiem, D., Kohli, P., Fergus, R.: Indoor segmentation and support inference from RGBD images. In: Fitzgibbon, A.W., Lazebnik, S., Perona, P., Sato, Y., Schmid, C. (eds.) ECCV (5). Lecture Notes in Computer Science, vol. 7576, pp. 746–760. Springer (2012)
30. Yang, J., Price, B.L., Cohen, S., Lee, H., Yang, M.-H.: Object contour detection with a fully convolutional encoder-decoder network. In: CVPR, pp. 193–202. IEEE Computer Society (2016)
31. Zitnick, C.L., Dollár, P.: Edge boxes: locating object proposals from edges. In: Fleet, D.J., Pajdla, T., Schiele, B., Tuytelaars, T. (eds.) ECCV (5). Lecture Notes in Computer Science, vol. 8693, pp. 391–405. Springer (2014)

Printed in the United States
By Bookmasters